AIDS
SCIENCE AND SOCIETY

Related Titles from Jones and Bartlett Publishers

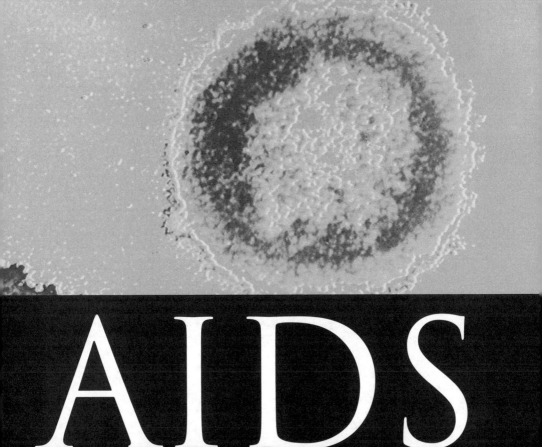

AIDS

SCIENCE AND SOCIETY

SIXTH EDITION

HUNG Y. FAN
ROSS F. CONNER
LUIS P. VILLARREAL

University of California, Irvine

JONES AND BARTLETT PUBLISHERS

Sudbury, Massachusetts

BOSTON   TORONTO   LONDON   SINGAPORE

World Headquarters

Jones and Bartlett Publishers
40 Tall Pine Drive
Sudbury, MA 01776
978-443-5000
info@jbpub.com
www.jbpub.com

Jones and Bartlett Publishers
Canada
6339 Ormindale Way
Mississauga, Ontario L5V 1J2
Canada

Jones and Bartlett Publishers
International
Barb House, Barb Mews
London W6 7PA
UK

Jones and Bartlett's books and products are available through most bookstores and online booksellers. To contact Jones and Bartlett Publishers directly, call 800-832-0034, fax 978-443-8000, or visit our website, www.jbpub.com.

Production Credits
Chief Executive Officer: Clayton Jones
Chief Operating Officer: Don W. Jones, Jr.
President, Higher Education and Professional Publishing: Robert W. Holland, Jr.
V.P., Sales: William J. Kane
V.P., Design and Production: Anne Spencer
V.P., Manufacturing and Inventory Control: Therese Connell
Publisher, Higher Education: Cathleen Sether
Acquisitions Editor: Molly Steinbach
Senior Editorial Assistant: Jessica S. Acox
Editorial Assistant: Caroline Perry
Senior Marketing Manager: Andrea DeFronzo
Text Design: Anne Spencer
Cover Design: Kristen E. Parker
Production Services: Mike Boblitt
Photo Research and Permission Manager: Kimberly Potvin
Associate Photo Researcher: Sarah Cebulski
Composition: Shawn Girsberger
Cover Images: Human immunodeficiency virus (HIV) under high magnification © Photodisc; AIDS quilt in Washington, D.C. © Mark Burnett/Alamy Images
Printing and Binding: Malloy, Inc.
Cover Printing: Malloy, Inc

Library of Congress Cataloging-in-Publication Data
Fan, Hung, 1947-
AIDS : science and society / Hung Y. Fan, Ross F. Conner, and Luis P. Villarreal. — 6th ed.
    p. cm.
Includes index.
ISBN-13: 978-0-7637-7315-1
ISBN-10: 0-7637-7315-8
1. AIDS (Disease) 2. AIDS (Disease)—Social aspects. I. Conner, Ross F. II. Villarreal, Luis P. III. Title.
RA643.8.F36 2011
614.5'99392—dc22
                        2009048744
6048

Printed in the United States of America
14 13 12 11 10   10 9 8 7 6 5 4 3 2 1

To our HIV-infected friends and acquaintances who are courageously battling the disease or who have succumbed to it. In their honor, and to hasten the day when this book is no longer necessary, a portion of the royalties from this book will be donated to foundations and community organizations dedicated to AIDS research and service.

Brief Contents

CONTENTS

Preface

THE PURPOSE OF THIS TEXT is to provide the nonspecialized student with a firm overview of AIDS from biomedical and psychosocial perspectives. The biological aspects include cellular and molecular descriptions of the immune system and the AIDS virus (human immunodeficiency virus, or HIV). The consequences of HIV infection from cell to organism are also covered, along with a clinical description of the disease. We then move from the organism level to the interorganism level covering both the psychological and social aspects of HIV/AIDS. These topics can be covered only in a survey fashion due to the comprehensive nature of this approach and the additional aim of making this text appropriate for a one-quarter (or semester) course (or part of such a course). We focus on presenting the relevant fundamental principles. Following a brief presentation of these principles for each topic, we generalize and apply these concepts to the case of AIDS.

This book began as *The Biology of AIDS*. The first edition of *AIDS: Science and Society* expanded from *The Biology of AIDS* and included consideration of social issues related to HIV/AIDS: personal risk assessment, HIV prevention, and the human and societal dimensions of living with HIV/AIDS. In the current edition of *AIDS: Science and Society*, we have provided updates and new sections in several of the chapters and provided the latest statistics on AIDS that were available as the book went to press. The book contains a reference appendix in which students can obtain additional information on AIDS. A major feature of this appendix is the inclusion of several organizations that have websites students can explore to find up-to-date information about AIDS.

This book is patterned after a one-quarter course, AIDS Fundamentals, taught at the University of California, Irvine. Approximately half the course covers biomedical aspects of AIDS, and the other half covers social issues raised by the disease. The text represents the material covered in the course. At UCI, AIDS Fundamentals is open to all undergraduate students and is taught with the assumption that they have had a high school level modern biology course. The material contained in Chapters 3 (immunology), 4 (virology), 6 (epidemiology), 9 (preventing HIV), and 10 (living with AIDS) is covered in three hours of lecture per chapter. Material covered in the other chapters is taught in a single one and one-half hour lecture per chapter. We have found that students are able to assimilate and retain the material when delivered at this rate. The course includes another important component: small discussion groups led by students who previously took the class. These peer-led groups use experiential exercises as a catalyst for a deeper understanding of the human and social aspects of HIV/AIDS. More details about this aspect of the course are in the instructor's resources available through the publisher. Another important feature of

the AIDS Fundamentals course is two panel presentations by people affected by HIV/
AIDS: a panel of people living with AIDS and a panel of HIV/AIDS healthcare workers.

Most researchers and scholars in AIDS-related fields were unprepared for the
dramatic impact of the AIDS epidemic when it emerged in 1981. As virologists and
social scientists, we might have expected modern biomedical technology to provide
a quick technical solution or to at least prevent, through vaccine development,
the spread of this major new viral epidemic. Although there has been biomedical
progress, it is now clear that the HIV/AIDS issues pose new and unforeseen difficulties
with no quick biological solution in sight. These difficulties challenge both our
scientific abilities and the ability of our society to respond appropriately. It is our
goal to provide students with a conceptual framework of the issues raised by HIV/
AIDS so that they will be better able to deal with the challenges posed by this disease.
This is particularly important because new information about scientific aspects of
HIV/AIDS appears regularly; with this information comes new implications for the
clinical, social, psychological, legal, and ethical aspects of the disease. We hope
that the framework provided in this book will help students understand and make
informed decisions about HIV/AIDS-related issues as they develop in the future.

Ancillaries

For Students
Jones and Bartlett Publishers has developed a *Companion Web Site* especially for
this text. Please visit **http://biology.jbpub.com/Fan/AIDS/6e** for useful study tools,
including summaries of the main points from each chapter, short-answer review
questions, and additional links to HIV/AIDS-related sites. Links to the organizations
mentioned in the appendix can also be found here.

For Instructors
The following materials are available for download:

- The *PowerPoint® Image Bank* provides all of the illustrations, photographs,
  and tables (to which Jones and Bartlett Publishers holds the copyright or has
  permission to reprint digitally) inserted into PowerPoint slides. With the
  Microsoft® PowerPoint program, you can quickly and easily copy individual
  image slides into your existing lecture slides.
- A set of *PowerPoint Lecture Outline Slides* provides outline summaries and
  relevant images for each chapter of *AIDS: Science and Society, Sixth Edition*.
  Instructors with the Microsoft PowerPoint software can customize the outlines, figures, and order of presentation.

For more information about these resources, please visit http://www.jbpub.com/
catalog/9780763773151.

Acknowledgments

We thank David Fan, Elaine Vaughan, Michael Gorman, David Prescott, Cedric Davern, David Baltimore, and Frank Lilly for reading parts of the original manuscript and providing many helpful substantive and editorial comments. Kathryn Radke and Ian Trowbridge provided helpful suggestions for a past revision; Ian also provided some of the content for the website linked to this book. Special thanks to the following reviewers who provided comments on previous revisions:

Robert Fullilove, Columbia University
James D. Haynes, State University College at Buffalo
James Rothenberger, University of Minnesota
Ian Trowbridge, Salk Institute
Thomas C. Van Cott, Henry M. Jackson Foundation

Juan Moreno applied outstanding computer graphic skills in generating some of the line drawings for the book. Bob Settineri of Sierra Productions was of great help in obtaining the other figures. Mike Boblitt, Cathleen Sether, Molly Steinbach, Jessica Acox, Caroline Perry, Andrea DeFronzo, and other staff of Jones and Bartlett were responsible for production of the final volume. We are grateful for their assistance and gentle prodding. We also thank Michael Feldman and Emmett Carlson for their love and support.

The Authors

Dr. Hung Fan is Professor of Virology in the Department of Molecular Biology and Biochemistry at the University of California, Irvine and Director of the UCI Cancer Research Institute. His research interest is in how retroviruses cause disease and induce cancer and AIDS.

Dr. Ross Conner is Professor Emeritus, School of Social Ecology, at the University of California Irvine, where he founded and directed the Center for Community Health Research at UCI. Dr. Conner continues his research in the area of community health promotion and disease prevention, including HIV, working in partnership with communities of many types and sizes in the United States and abroad. His work also includes the evaluation of the effectiveness of social programs and public policies. He is the president of the International Organisation for Cooperation in Evaluation.

Dr. Luis Villarreal is Professor of Virology in the Department of Molecular Biology and Biochemistry at the University of California, Irvine. Dr. Villarreal's research interest is in the strategy of how viruses replicate and how they cause disease.

CHAPTER 1

Introduction: An Overview of AIDS

A report appeared in 1981 that initially drew little attention from infectious disease experts. In that report, Dr. Michael Gottlieb, at the University of California at Los Angeles, described a rare form of pneumonia occurring in homosexual men. Other reports from about the same time indicated that other homosexual men were developing a rare form of cancer. This new set of symptoms, a *syndrome* in medical terms, was eventually called *acquired immune deficiency syndrome* because the symptoms were consistent with damage to the immune system in previously healthy individuals. Moreover, this disease was not congenital or inherited but appeared to have been acquired. (We now know it results from infection by a virus.) Since then, the acronym *AIDS*, which is used to describe this disease, has become a prominent and permanent fixture in our language. It evokes a range of responses, including fear, hate, and mistrust. Some of these responses (hate, mistrust) are related to the association of AIDS with subcultural groups within our society, such as male homosexuals, who already have experienced discrimination. Other responses (fear) are due to the grave nature of the AIDS disease and the threat it may pose to society. This fear is because the AIDS epidemic continues to grow—unlike most other major infectious diseases, which have been controlled by a combination of clinical treatments and public health measures.

AIDS in Brief

We now know that AIDS is caused by the *human immunodeficiency virus (HIV)*, but it was originally observed by its effects on the immune system. An important clue was that AIDS patients often developed a lung infection (or pneumonia) caused by a fungus called *Pneumocystis*. This infection is very rare in healthy individuals, but patients with cancers of the immune system (lymphomas) were known to be susceptible to this disease. Lymphomas are usually treated by chemotherapy, which is intended to destroy the cancer cells. However, chemotherapy also unavoidably destroys many healthy immune cells along with the cancerous lymphoma cells. Thus, this type of pneumonia predominantly occurs in patients with damaged immune systems. Examination of AIDS patients confirmed that their immune systems were damaged. The specific nature of this damage is discussed in greater detail in Chapters 3 and 4. It had been known for some time that various other viral infections could damage cells of the immune system, but the severe damage seen with AIDS was unprecedented. Although doctors suspected early on that AIDS resulted from infection by a virus, it was not until 1984 that the virus (HIV) was finally isolated by both French and American researchers.

In addition to pneumonia, AIDS is associated with numerous other infections. These secondary infections are caused by various bacteria, protozoa, fungi, and other viruses. Usually, it is the secondary infection(s) (known as an *opportunistic infection*) that causes death in AIDS patients. In addition to secondary infections, AIDS patients frequently develop cancers, including *lymphomas* and an otherwise rare cancer called Kaposi's sarcoma. HIV infection also can result in damage to brain cells. This damage leads to loss of mental function, referred to as AIDS *dementia*. A more complete description of the clinical features of AIDS is presented in Chapter 5. Most of these opportunistic infections and some other effects of HIV infection can be explained by damage to the immune system.

HIV causes disease insidiously. The early stages of infection may not be noticed by the infected individual. The infected person may feel healthy and appear to be completely normal during this time (the asymptomatic period), but such a person is able to transmit the infection. The HIV *incubation period* (the time between initial infection and appearance of disease) is of variable duration and can be quite long (on average, 10 years or more). In contrast, for most common viral infections, such as colds or influenza, an incubation period of a few days or weeks is followed by apparent disease. This adds greatly to the difficulty of studying and controlling AIDS, because many people infected with the virus have not yet developed the disease.

The AIDS Epidemic

Despite the many different clinical symptoms that result from AIDS, medical investigators know a great deal about how AIDS is spread in our population. For example, it is now clear that HIV transmission requires close contact and that infection occurs by one of three routes: blood, birth, or sex. Casual contact does not lead to disease transmission. These issues are further discussed in Chapter 7.

Between 1981 (the beginning of the AIDS epidemic) and 2007, a total of 1,051,875 AIDS cases in the United States were reported to the national Centers for Disease Control and Prevention in Atlanta, Georgia. Of these cases, about 583,298 (55%)

have died. Sexually active homosexual males were originally the major afflicted group and currently represent about 46% of these reported cases. Another 24% of the cases are male or female injection drug users, and 7% are male homosexual drug users. Another 23% result from heterosexual transmission, birth, or blood transfusion during the period when the American blood supply was not monitored for HIV antibodies (1981–1985).

In the relatively brief period since the beginning of the AIDS epidemic, AIDS has already had a major impact on death and disease in the United States. Currently, there are between 40,000 and 60,000 new cases of HIV infection every year, and the number of people dying from AIDS per year is currently approximately 15,000. In comparison, approximately 40,000 women die each year from breast cancer, and about 35,000 men die each year from prostate cancer. On the other hand, the average age of death from breast or prostate cancer is considerably older than for death from AIDS. The AIDS epidemic has had a particularly high impact on African Americans and Hispanics, who show rates of HIV infection that are three to six times higher than that of the general population.

The AIDS epidemic is not restricted to the United States. It can be found on all continents and hence is considered a *pandemic*. It is estimated that 22 million people in sub-Saharan Africa are infected with HIV. In Africa, HIV transmission predominantly results from heterosexual contact and other modes. Given the relatively poor medical support available in much of Africa, the number of deaths from AIDS will increase significantly. There are high levels of HIV infection in certain countries of Asia, and it is spreading explosively in some parts of eastern Europe. Because there is no cure for AIDS, these numbers are alarming. They indicate the clear potential of HIV/AIDS to spread unchecked, despite recent advances in modern medicine, epidemiology, virology, and recombinant DNA technology. This threat reminds us of earlier times when major infectious diseases devastated human populations (see Chapter 2).

Worldwide, AIDS now ranks as the fourth leading cause of death after heart disease, stroke, and acute lower respiratory infections. In Africa, it is the leading cause of death. How can we control this epidemic? An overview of the relationship between epidemics and human populations may shed some light on this concern.

http://biology.jbpub.com/fan/aids/6e/

Connect to this book's website: http://biology.jbpub.com/fan/aids/6e/. The site features summaries of the main points from each chapter, links to important AIDS-related websites, and short-answer-style review questions for each chapter.

CHAPTER 2

Concepts of Infectious Disease and a History of Epidemics

One of the great recent achievements of modern civilization has been the control of infectious diseases. Many of us may not personally know anyone who died from a contagious disease. In historic terms, this is a new development, one that began in the mid-twentieth century. In previous centuries, death from infectious disease was common, and whole populations were often affected.

When a population becomes infected with a contagious disease, an epidemic results. *Epidemic* derives from Greek and means "in one place among the people." To understand how an infectious disease can spread or remain established in a population, we must consider the relationship between an infectious disease agent and its host population. The study of diseases in populations is an area of medicine known as *epidemiology* and is discussed further in Chapter 6.

Contagious diseases are spread by microorganisms, such as certain bacteria and viruses, that cause disease when they infect a susceptible person. This manner of disease transmission is a modern concept, known when it was developed as the *germ theory* of infectious disease. Before this understanding, earlier societies often used moral or religious explanations for infectious disease, and social practices developed that reflected those beliefs.

Factors That Affect the Spread of Epidemics

In this section, we will discuss factors that influence the spread of infectious diseases. Although many different microorganisms cause diseases, we will focus on viruses because HIV is a virus. The general principles are the same for other infectious microorganisms.

Host and Virus Populations

An epidemic consists of infection of a number of individuals in a population. It is important to look at more than a single person to understand how diseases spread. Two populations must be considered: the human host and the infecting agent—in the case of AIDS, a virus. These two populations have a balanced host–parasite relationship. A viral infection can deplete or limit the population of its host, but a highly lethal virus that spreads too rapidly might kill all available hosts and lead to the extinction of both its host and itself. The course of an epidemic, however, is not always straightforward. It can be influenced by a number of other factors about the population:

1. The total number of hosts
2. Their birth rate
3. The rate at which susceptible individuals migrate into the population
4. The number of susceptible hosts who are not infected
5. The rate at which the disease can be transmitted from an infected individual to an uninfected one
6. The number of infected individuals who die
7. The number who survive the infection and become immune or resistant to further infection

Figure 2-1 shows a schematic relationship of people in a population with regard to a simple viral infection; all infected individuals either recover from the infection and become immune to it, or they die from it. The population can be divided into those who have not been infected by the virus (susceptibles), those who are infected, and those

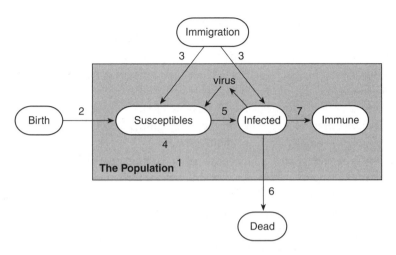

Figure 2-1   Population factors that affect epidemics: (1) population size, (2) birth rate, (3) immigration rate, (4) number of susceptibles, (5) transmission rate, (6) death rate, and (7) immune rate.

who have recovered from the infection and are immune to it. The arrows that connect the boxed groups represent movement of people from one group to an adjacent one. New susceptibles enter the population by birth or immigration. Infected people are the source of the virus that infects new susceptibles; they either result from infection of susceptibles or immigration.

This scheme is a simplified representation of the dynamics or ecology of a virus epidemic. It is possible to develop mathematical models to describe or predict the course of an epidemic if the rates of movement through the scheme can be determined. One of the applications of the field of epidemiology (see Chapter 6) is to determine these rates.

The Transmission Rate

The arrow in Figure 2-1 that connects the susceptibles to the infected group (labeled 5) is the *transmission rate* of infection. This rate represents the efficiency with which disease is transmitted from an infected person to a susceptible person. This transmission rate has two major components (Figure 2-2). One is the *inherent efficiency* with which

Figure 2-2   Transmission rate of infections (factor 5 in Figure 2-1) has two major components: (1) inherent efficiency of virus infection and (2) encounter rate between infected and uninfected.

a particular virus can infect a susceptible person. The inherent efficiency of infection for a virus is dependent on the biological properties of the virus and the route by which the virus enters the susceptible person. For example, influenza virus, like many other viruses that infect the respiratory tract, has a high inherent efficiency of infection and is highly contagious. Respiratory viruses are easily taken up by breathing in aerosols (e.g., sneezes); once influenza virus comes into contact with cells of the respiratory tract, it readily infects them. HIV, on the other hand, actually has a relatively poor inherent infection efficiency, as we shall see later.

It should be noted that if an infectious agent has more than one mode of infection, the inherent efficiency of infection may be different for each mode. An extreme example of this is the bacterium that causes the plague (see later in this chapter). The plague bacterium is typically spread by flea bites, resulting in the bubonic form of plague—in bubonic plague, infection mostly involves the lymph nodes inside the body. On the other hand, if the same bacterium infects the lungs, it can be directly transmitted from infected to uninfected people by coughing (the pneumonic form); this pattern of spread has a much higher inherent efficiency of infection.

The other major component of the transmission rate is the rate at which a susceptible person in the population encounters an infectious person—the encounter rate. Each encounter between an infected person and an uninfected person increases the likelihood that an infection will be transmitted.

As we shall see later with the AIDS virus, both of these components of transmission can be changed by altering the behavior of susceptible and infected persons. Behaviors that allow high encounter rates with infected people or that allow more efficient infection will favor the spread of an epidemic. Conversely, changes in behavior that reduce these transmission factors may control the spread of an epidemic.

Population Densities and Infections

Many of the epidemics that have plagued humankind for the last few thousand years would not have had a favorable transmission rate during early human civilization. Early human societies were not urban but consisted of hunter–gatherers who lived in relatively small groups such as extended families. Such small groups or populations may not be able to produce new susceptibles in high enough numbers at any given time to support the continued growth of disease-causing microorganisms. An acute disease produces symptoms and makes a person infectious soon after infection. The infected person transmits the disease, dies from the infection, or recovers and becomes immune to subsequent infections. An acute microorganism that strikes such small groups quickly infects all available susceptibles and then dies out.

About 10,000 years ago, the agricultural revolution allowed human populations to become large enough to support epidemics. In other words, the development of complex human societies was necessary before epidemics by acute viruses occurred. When the world population became sufficiently large, different patterns of infection also could develop. Acute epidemic diseases and agents could establish an *endemic pattern*—one in which the infectious disease is always present in some members of the large population. After the initial introduction and spread into a susceptible or naïve population, even a very lethal virus can become endemic. For endemic viruses, the numbers who are actively infected in the population are much lower, but they are always present. They serve as a source of infection for susceptibles. Endemic viral infections are often considered childhood diseases because the virus is so common in the population that most individuals encounter it during childhood. Most adults have been infected (with or without symptoms), and they have survived and are immune. Endemic viral infections may also have contributed to the high infant mortality in previous times (e.g., Europe in the Dark and Middle Ages) and in some developing countries today.

Endemic infectious agents can limit population sizes and result in populations that are relatively unaffected or are resistant to the infectious agent as a whole. As we shall discuss, when two previously separated societies encounter each other for the first time, the introduction of an endemic infectious agent into a population that has not previously encountered it (a *naïve* population) can have drastic results.

Chronic Infections

In addition to acute infections, such as measles, there are chronic infections. In *acute infections*, the disease symptoms generally occur quite soon after infection, and the infectious agent is generally eliminated by the individual's immune system after the initial disease period. Some people infected with an acute virus do not develop symptoms (subclinical infections); nevertheless, they will generally eliminate the virus. In a *chronic infection*, the person does not eradicate the infectious agent (often a virus). The virus persists in the infected person, sometimes at low levels. People with chronic infections often do not show symptoms or disease immediately after infection. The differences between acute infection and chronic infection in an infected individual are diagrammed in Figure 2-3. As described earlier, acute infections generally require large populations (with continued new susceptibles) to be maintained. In contrast, chronic infections can sometimes be maintained in small populations. In addition, chronic infections are often more difficult to control at the population level because infected and uninfected people may be indistinguishable. Moreover, in persistent infections, the infected individual has not mounted an effective immune response to eliminate the virus. As we shall see, the syphilis epidemic was difficult to control partly because it is a chronic infection. Like syphilis, AIDS results from a chronic infection.

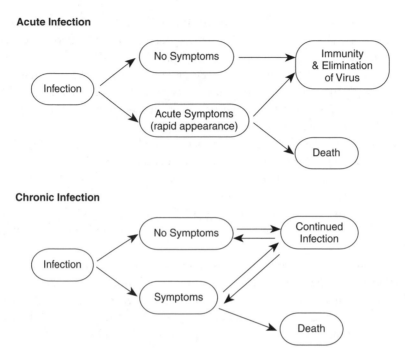

Figure 2-3   Acute versus chronic infection. These charts present the consequences of infection by an acute virus compared with infection by a chronic virus. The frequencies with which death, immunity, or continued infection occur are different for different viruses.

Controlling Infectious Diseases

Since the beginning of the twentieth century, there has been a steady and dramatic decrease in the number of people who die from infectious diseases, particularly in developed countries. Recently, most developed countries have been free of major lethal contagious diseases. *Antibiotics* can kill bacterial infections after they start. Viruses pose a different problem: They are difficult to eliminate once they become established. Therefore, viral diseases have been controlled mostly by vaccination (see Chapter 3, p. 33) but occasionally by other measures. A vaccine induces immunity to a virus in susceptible individuals without the individuals becoming infected (Figure 2-4). Because birth is the major source of susceptibles in a population, in many cases, newborns are vaccinated. In fact, if enough (but not necessarily all) susceptibles are immunized, this can confer immunity on the population as a whole. This is because the remaining unimmunized but susceptible individuals are unlikely to encounter another infectious individual. Thus it is possible to completely eliminate some infectious diseases from the human population with an effective vaccination program, even if not everyone is vaccinated. The smallpox virus, which was responsible for so much human death in

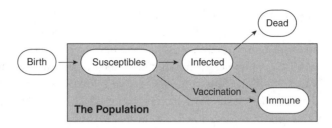

Figure 2-4   Epidemic control by vaccination.

historic times, is now eradicated from the general population because of successful worldwide vaccination efforts.

A History of Epidemics
The Old World
Even the very earliest historic records document the major impact of epidemics. It is not always clear to us now which infectious agent was causing a particular epidemic in ancient times, but we can often make guesses from the recorded symptoms. The three disease agents that have probably caused most human deaths are smallpox virus, measles virus, and plague bacterium (*Yersinia pestis*). These three diseases have accounted for hundreds of millions of human deaths over the years and an unfathomable amount of human suffering. Other important epidemic agents include influenza virus, typhoid fever bacterium, yellow fever virus, polio virus, and, more recently, hepatitis viruses. The syphilis bacterium *Treponema pallidum* is of special interest here because of its sexual mode of transmission and its associated social problems.

Many historic accounts make clear reference to a supposed religious or moral reason for a particular epidemic. The transmission of disease itself was often believed to occur through casting of an evil eye. In the book of Exodus in the Old Testament, for example, God punished the Egyptian pharaoh for enslaving the Israelites by bringing upon Egypt a plague of "sores that break into pustules." Epidemics were often perceived as punishment due to the wrath of a deity, perhaps for some offense by the entire population. Those who developed a disease were viewed as deserving it. This tendency to link a disease to social stigmatism has persisted throughout history and afflicts people with AIDS today.

The Greek writings are probably the earliest accounts in sufficient detail to allow us to measure the impact of epidemic disease. Aside from malaria, the Greeks were relatively free of most infectious diseases with one important exception. In 430–429 BC, an epidemic that may have been measles struck Athens with a devastating loss of life. It also resulted in a significant decrease in the size of its armies, and the following year Athens lost a war with Sparta. Thus, this epidemic may have influenced history.

The Roman Empire also suffered massive epidemics in 165 AD and again in 251 AD. Before the 165 AD epidemic, the population of the Roman Empire was probably at its peak (about 54 million). After the 165 AD and 251 AD epidemics, the Roman population did not recover its size until modern times. The first epidemic could have been smallpox and appears to have killed one-third of Rome's population. The 251 AD epidemic may have been measles and was equally devastating. There were about 5,000 deaths per day in Rome at the epidemic's peak. Rome's rural population may have been even more affected. This die-off may have led to the depopulation of agricultural lands and an inability to oppose invasion from the north. A third massive epidemic occurred in 542–543 AD, probably due to bubonic plague. Soon after this plague, Rome's armies fell to the Visigoth and then to the Muslim armies, and the Dark Ages of Europe began. Thus, epidemiological history suggests that infectious diseases may have contributed to the fall of the Roman Empire.

The situation for China, although more difficult to estimate, appears to have been similar. Massive epidemics in 162 AD and again in 310 AD may account for much of the population decline in China, which peaked at about 50 million at those times but declined to about 8.9 million by 742 AD.

In Europe, and probably also in China, measles and smallpox eventually became endemic childhood diseases after these devastating epidemics. In the following millennia, Europe experienced devastating epidemics from the disease known as the *black death*. Black death was a pneumonic (or lung infection) form of plague, which had a very high fatality rate. It probably accounted for up to 100 million deaths in Europe. The worst of these epidemics occurred in 1346. This epidemic appears to have been a *pandemic*, meaning that other continents (China and India) also were involved. The black death recurred in Europe in the 1360s and again in the 1370s. The seemingly arbitrary pattern of death and the massive suffering had dark social consequences for Europe. *Xenophobia*, the fear of foreigners, became common. Violent riots against Jews and Gypsies occurred in numerous cities because they were blamed as sources of the plague. Self-flagellation became a common practice, and rational theology lost popular acceptance. The situation improved somewhat in the 1400s. Black death became endemic, possibly because of selection for a less virulent plague bacterium; selection for people with greater resistance to the disease also may have occurred. European society was now experiencing most of these acute infectious diseases, especially the viral diseases, as childhood diseases.

The New World

A well-documented example of what happens when a new viral disease enters a naïve population (one that has never encountered the virus) occurred when Spanish conquistador Hernán Cortés went to Mexico and introduced smallpox into the New World. The Aztec Codices (hieroglyphic-like records) tell us that the New World was relatively free of major infectious disease at that time. The population of Mexico was probably 25–30 million, and Mexico City may then have been the most populous city in the world. In 1518, just as the Aztecs drove Cortés from Mexico City, a smallpox epidemic swept through the city, killing the Aztec leaders and decimating the city's population. This epidemic was followed by numerous other diseases that were European endemic childhood diseases but were devastating to the Aztecs. Within 50 years, the population of Mexico was down to about 1.5 million, or about 5% of what it had been at its peak. Furthermore, the fact that the diseases seemed to strike only the Aztecs and not the Spaniards led the Aztecs to believe that the gods favored the Spaniards.

Other American natives fared even worse than the Aztecs: The Native Americans of Baja, California, and other island tribes became totally extinct. Thus, the main fabric of native American society was utterly destroyed. Mexico began to recover from this population loss only in the 1800s, and only now has Mexico City become the most populous city in the world again. A similar fate was in store for the Pacific island natives, who also suffered huge population losses after encountering European explorers. Thus, throughout human history, infectious diseases have profoundly affected human populations.

Modern Concepts of Infectious Disease and Koch's Postulates

The germ theory of disease—the idea that a microorganism or "germ" causes an infectious disease—was first proposed in 1546 by Girolamo Fracastoro, a Franciscan monk. However, it was not until the 1840s that Friedrich Henle, a German physician, clarified these concepts and they became accepted among scientists. One of Henle's students, Robert Koch, subsequently proposed four postulates that could be used to prove that an infectious agent causes a disease. This was a milestone in the understanding of infectious disease. *Koch's postulates* state that an organism can be considered to cause a disease if it fulfills the following criteria:

1. The organism is always found in diseased individuals.
2. The organism can be isolated from a diseased individual and grown pure in culture.
3. The pure culture will initiate and reproduce the disease when introduced back into a susceptible host (either human or animal).
4. The organism can be reisolated from that diseased individual.

These postulates allowed scientifically sound assignments of what agents caused specific diseases, and they freed physicians from many superstitions and myths that had historically prevailed.

Actually, by today's standards, Koch's postulates are sometimes too stringent. For example, viruses cannot be grown pure in culture in the absence of cells (see Chapter 4). Also, if two infectious agents cooperate to cause a disease or a particular set of symptoms, it also would be impossible to fulfill Koch's postulates. We shall see that this situation applies to HIV infection and AIDS. In the late 1800s, however, such stringency was necessary.

The timing of the development of Koch's postulates, and of the development of the science of epidemiology, was most fortunate because other changes in society set the stage for the outbreak of another worldwide pandemic. In the late 1800s, steamships brought about relatively rapid world travel. This change had an impact on the ecology of infectious disease by allowing the rapid movement of infected persons who could quickly spread an epidemic. In 1894, another plague pandemic broke out, initially in Burma, then in Hong Kong, and then via steamships to all major ports worldwide, including those in the United States. By applying the germ theory of disease and epidemiology, society was able to respond to this threat. The application of Koch's postulates led to the rapid identification and isolation of the causative bacterium, *Yersinia pestis*. Furthermore, intense epidemiological studies identified rats, and more specifically their fleas, as major carriers of the disease. This finding led to the development of preventive measures to control the spread of the plague, principally by limiting interactions between rats and humans. Except for a further breakout in India, the plague pandemic was stopped. This was an important lesson. There was no cure or vaccine for plague at that time, yet understanding the routes of infection and designing measures to minimize spread of infection based on this understanding averted a pandemic. With the current AIDS epidemic, we are in a similar situation because there is no cure or vaccine. Behavior modification to minimize the spread of the AIDS virus is currently our primary means of controlling the epidemic. However, because AIDS is predominantly a sexually transmitted disease, behavior modification is difficult.

Epidemics in Modern Times

In the twentieth century, several other epidemics took a toll on humanity. During the great pandemic of 1918, influenza virus killed about 20 million people worldwide and virtually brought World War I to a halt. About 80% of American casualties in World War I were caused by influenza, a fact seldom mentioned in most history texts. Influenza continues to cause epidemics and remains a health threat. The major reason is that this virus can mutate rapidly. These mutations lead to changes in the surface structure of the virus that allow the virus to avoid the immune system. As a result, individuals who were previously infected with influenza virus are not protected from the new mutant virus. As we will see later, HIV also has a similar property.

Recently there has been concern about two new influenza epidemics. Beginning in 2003, an epidemic of avian (H5N1) influenza has infected domestic and wild birds in Asia and Europe. In Asia, rare cases of transmission of this virus from birds to humans have occurred (about 200 cases in the past 10 years), but the death rate in infected humans is very high—approximately 50%. While there have been few cases of H5N1 influenza in humans so far, public health authorities are worried that the current H5N1 avian influenza might mutate to spread efficiently between humans. Such a mutation could set the stage for an influenza pandemic that rivals the Spanish influenza of 1918. In March 2009, a new influenza in humans appeared, first in Mexico, and it has now spread worldwide. Thus it is a new pandemic. This virus, called H1N1, is a combination of human influenza, swine influenza, and avian influenza. It is sometimes also called "swine flu" or "atypical H1N1" influenza. The severity of the new H1N1 influenza is not clear yet, although there have been deaths associated with the infection.

Poliovirus is another recent epidemic disease. This disease appeared as a new viral epidemic in the United States in 1894—much as the AIDS epidemic appeared in 1981. Poliovirus can damage the nervous system and lead to paralysis. In contrast to HIV, which entered North America in the late 1970s (see Chapter 4), poliovirus had been infecting people since early history, but it did not cause documented epidemics until 1894. Paradoxically, improvements in hygiene and sanitation in more developed societies actually predisposed individuals to the paralytic form of polio by delaying exposure to the virus until they were young adults. Infection of infants, which tends to occur in less developed countries, usually results in a mild nonparalytic gastrointestinal infection. Thus, the people most likely to get paralytic polio were healthy young adults from the highest socioeconomic classes. Polio had a major impact on the American consciousness, as seen by highly visible national crusades during the first half of the twentieth century (such as the March of Dimes). This underlines the way in which the nature of the victims can influence public perceptions of a disease and society's response to it. In fact, there were about 50,000 total deaths from paralytic polio during the first half of the 20th century. It is interesting to contrast the public response to polio during this time to recent responses to AIDS—even though more deaths from AIDS occurred in the United States in the first 10 years of the epidemic.

Syphilis: The Social Problems with a Sexually Transmitted Disease

One epidemic that is hauntingly similar to the AIDS epidemic is syphilis. The parallels are striking. At the time of the syphilis epidemic, scientific investigation of this insidious disease was at the leading edge of medicine and microbiology, as is the current situation with AIDS. The issues raised included public health policy and civil liberties, again as in the AIDS epidemic. And finally, because syphilis is a sexually transmitted disease, patients with that disease were highly stigmatized. A cure for syphilis in infected individuals was developed in 1909, but it was not until the 1940s that the epidemic was finally controlled.

Why did it take so long to control this epidemic? Like AIDS, syphilis can be a long-term and variable disease, with phases in which no symptoms are apparent. Unfortunately, untreated syphilis often eventually leads to death. More important, at that time syphilis was perceived as a social problem—hence the reference to it as a *social disease*. Many blamed the disease on a breakdown of social values and promoted the view that a sexual ethic in which all sex was marital and monogamous would make it impossible to acquire the disease. The initial public health policies to control this epidemic were based on these views. Abstinence from extramarital sexual contact was encouraged, and prostitution was repressed because prostitutes were blamed as the major source of infection of otherwise monogamous males. Immigrants were also blamed for bringing the disease from abroad, even though epidemiological data did not support this view. As many as 20,000 prostitutes were *quarantined* or jailed during World War I. In addition, the Army discouraged the availability of condoms for fear that they might encourage soldiers to engage in extramarital sex. There were also campaigns to stigmatize soldiers who became infected with syphilis by giving them dishonorable discharges. These policies were not based on epidemiological evidence, and they failed to control the epidemic, which actually grew during this period.

It was not until the 1930s that the surgeon general of the United States, Thomas Parran, proposed major changes in the public health approaches to control the syphilis epidemic. These policy changes were ultimately successful but required substantial funding from Congress. The proposals called for the elimination of repressive approaches that discouraged people from participating in programs or seeking treatment. Free and confidential diagnostic and treatment centers were set up throughout the nation. A national educational campaign was begun to educate the public and dispel prevalent misconceptions (even among respected sources) about the disease's transmission. Syphilis is transmitted by sexual contact but not by casual contact. These policies, along with new antibiotics, brought the syphilis epidemic under control in the 1940s.

With the AIDS epidemic, we are dealing with powerful biological drives such as human sexuality and drug addiction. The syphilis epidemic shows us that policies based mainly on abstinence are not very effective in controlling a sexually transmitted disease. Other alterations in behavior are necessary to reduce the transmission of AIDS and bring this epidemic under control. Until a cure or a vaccine against AIDS is developed, changing behavior (see Chapter 9) is our most effective means of controlling the AIDS pandemic.

http://biology.jbpub.com/fan/aids/6e/

Connect to this book's website: http://biology.jbpub.com/fan/aids/6e/. The site features summaries of the main points from each chapter, links to important AIDS-related websites, and short-answer-style review questions for each chapter.

CHAPTER 3

The Immune System

As mentioned in Chapter 1, AIDS results from a viral infection that ultimately disables the immune system. To understand this disease, we need to understand the immune system. This system is an intricate collection of cells and fluids in our body that gives us the ability to fight off infections. HIV, the AIDS virus, specifically affects certain cells of the immune system. Once we know about these cells and what they do, we can see how HIV does its damage. This chapter provides a simplified overview of immunity; many more intricacies and details are known, but the information provided here allows us to understand the basic immunological problems associated with AIDS.

Blood

To understand the immune system, we must first consider blood. Blood is a system of circulating cells and fluids that carries out many important functions for the body. These functions include transport of nutrients and oxygen to the body tissues, elimination of waste products and carbon dioxide from tissues, wound repair, and protection from infection by foreign agents. Besides cells, the fluid portion of blood contains many different substances and molecules that help to carry out these functions. Some examples are sugars, which are necessary for energy metabolism in our tissues, and antibodies, which are important in fighting infections. The cell-free fluid portion of blood is referred to as *plasma*. *Serum* can be obtained from isolated blood by letting it stand and clot; the cells are trapped in the clot and can be removed easily. Cells and substances of the blood that are responsible for protection from infection make up the *immune system*. The immune system must protect us from a wide variety of infectious agents, as follows (in ascending order of complexity):

Viruses: These are very small subcellular agents (see Chapter 4).

Bacteria: These are small single-cell microorganisms that have relatively simple genetic material. Typhoid fever and tuberculosis are caused by bacteria.

Protozoa: These are single-cell microorganisms that contain more complicated genetic structures. Amoebas and *Giardia* are examples of protozoa.

Fungi: These are more complex microorganisms that may exist as single cells, or they may be organized into simple multicellular organisms. Examples are yeasts and molds.

Multicellular parasites: These can be relatively large organisms, such as roundworms and tapeworms.

In addition, the immune system is also important in fighting cancer.

Blood is carried throughout the body by a series of blood vessels that make up the *circulatory system* (Figure 3-1). The heart is the pump for the circulatory system, and it moves blood through the blood vessels by its rhythmic muscular contractions. There are three kinds of blood vessels: *arteries*, which carry blood away from the heart to the body tissues; *veins*, which carry blood back to the heart from the tissues; and *capillaries*. Capillaries are very thin-walled blood vessels in the tissues that connect the arteries with the veins and allow exchange of oxygen, nutrients, and wastes between the blood and tissues. Some types of blood cells (such as the white blood cells, called *monocytes* and *lymphocytes*) can also pass through these thin walls from the blood into the tissues. The lungs are another important part of the circulatory system; it is here that the exchange of oxygen and carbon dioxide between the blood and the air we breathe takes place.

The cells in the blood have limited life spans—ranging from 1 or 2 days to several months, depending on the cell type. This means they must be continually replenished. They are replenished from *stem cells* that are mostly located in the bone marrow. These stem cells have the capacity to divide and make more of themselves and to differentiate and mature into blood cells of all types (Figure 3-2). During the differentiation

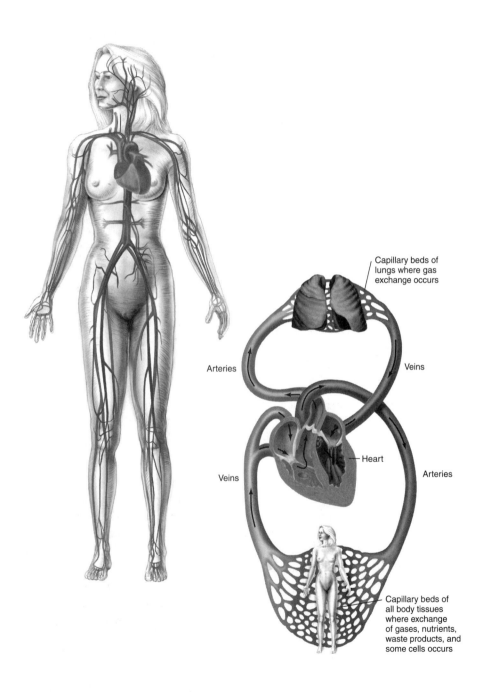

Capillary beds of
lungs where gas
exchange occurs

Arteries

Veins

Heart

Veins

Arteries

Capillary beds of
all body tissues
where exchange
of gases, nutrients,
waste products, and
some cells occurs

Figure 3-1   The circulatory system.

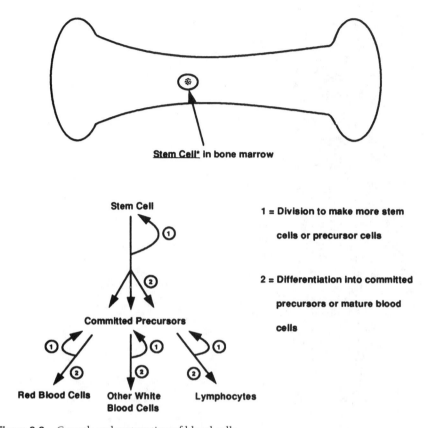

Figure 3-2 Growth and maturation of blood cells.

process, the stem cells first develop into *committed precursors*, which can either divide or differentiate into mature blood cells of a particular kind. This process goes on throughout life and, if interrupted, results in very serious health problems. When stem cells and committed precursors divide or differentiate, they require the presence of *growth factors* to carry out these processes. Different growth factors stimulate particular kinds of blood cells, and these growth factors play important roles in regulating the orderly growth and replenishment of all blood cells. For example, *interleukin-2* (IL-2) is a growth factor that is required by blood cells called T-lymphocytes, which are discussed later.

Cells of the Blood

Let us now look at the different kinds of cells present in blood. These cells are shown in Figure 3-3. Blood cells are divided into *red blood cells* and *white blood cells*. There are

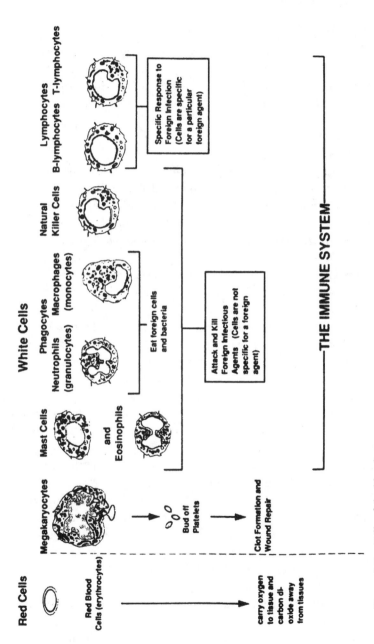

Figure 3-3   Cells of the blood.

a lot of red blood cells, but they are a single cell type; the white blood cells are fewer in number, but they are made up of many cell types.

Red Blood Cells

Red blood cells, or *erythrocytes*, are responsible for carrying oxygen to the tissues and carbon dioxide away from them. They contain a protein called *hemoglobin* that binds and carries the oxygen and carbon dioxide within them. Hemoglobin gives red blood cells their characteristic red color. All other blood cells are called white blood cells because they lack hemoglobin.

White Blood Cells

White blood cells, or *leukocytes*, are of several different types. *Megakaryocytes* are very large blood cells that actually do not circulate; they bud off subcellular fragments, called *platelets*. Platelets circulate through the bloodstream, and if they encounter a break in a blood vessel, they cause a clot to form. Thus, they are involved in wound repair.

Cells of the immune system are of two classes: those that respond to a specific foreign agent or substance and those that are *not* specific for the agent they attack. The cells that are specific for a certain foreign agent are *lymphocytes*. Cells that are not specific for the foreign agent they attack include *phagocytes*, *mast cells*, *eosinophils*, and *natural killer cells*.

Phagocytes are cells that attack and eliminate foreign cells or bacteria by engulfing or eating them. *Phagein* is the Greek word for "to eat." There are two different kinds of phagocytes: *macrophages* (and related monocytes) and *neutrophils* (also called *phagocytic granulocytes*). Macrophages generally attack and engulf cells infected with viruses, and neutrophils generally attack foreign bacteria. Macrophages are found not only in the blood (where the immature forms are called monocytes) but also in tissues.

Mast cells, basophils, and eosinophils attack infectious agents that are too large to be engulfed by a single blood cell. Such agents include protozoa and large parasites, such as worms. Mast cells and eosinophils come into contact with the foreign agents and release toxic compounds that may kill them.

What instructs phagocytes, mast cells, and eosinophils to attack a foreign agent? In general, *antibodies*, which are proteins produced by certain lymphocytes (see page 25), first bind to the foreign agent in a specific fashion. The phagocytes, mast cells, or eosinophils then recognize the agent because of the antibodies bound to it, and they attack. This process is diagrammed in Figure 3-4.

Lymphocytes are cells that respond *specifically* to particular foreign substances, or *antigens*. It is important to define an antigen. An antigen is a molecule or substance against which lymphocytes will raise a response. An example of an antigen is a protein of a virus particle (or even a small portion of a viral protein); the number of possible antigens that we might encounter is limitless.

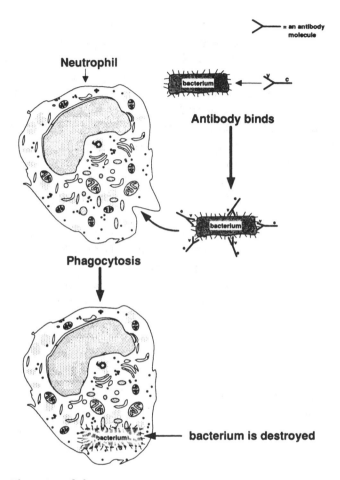

Figure 3-4   The action of phagocytes.

Lymphocytes are divided into two types: B-lymphocytes and T-lymphocytes. *B-lymphocytes* secrete soluble proteins called *antibodies* into the circulatory system. Each individual antibody specifically recognizes and binds to one particular antigen. Once this happens, the antibody signals other cells in the immune system to attack (Figure 3-4). In addition, certain antibodies, called *neutralizing antibodies*, may bind directly and inhibit the function of infectious agents such as viruses.

T-lymphocytes (or T-cells) make proteins called *receptors* that are similar to antibodies in that these proteins recognize specific antigens. However, T-lymphocytes do not release their receptors but hold them on their cell surfaces. As a result, the T-lymphocytes themselves specifically recognize and bind to foreign antigens.

The two major kinds of T-lymphocytes are *cytotoxic*, or *killer T-cells* (T_{killer}), and helper T-cells (T_{helper}). T_{killer} cells directly bind to cells carrying a foreign antigen. Once they bind to them, they attack and kill those cells, thus eliminating them from the body. T_{helper} cells, on the other hand, do not kill cells. Instead, they interact with B-lymphocytes or T_{killer} lymphocytes and help them respond to antigens (more about this later). In addition to the receptors, T_{killer} and T_{helper} cells each have characteristic proteins on their surfaces: the CD8 protein is present on T_{killer} cells, and the CD4 protein is present on T_{helper} cells (Figure 3-5). Simple tests have been devised for the CD4 and CD8 proteins, and they can be used to identify and count T_{killer} and T_{helper} lymphocytes.

T-lymphocytes get their name from the fact that their maturation depends on passage through the thymus gland. The thymus is a butterfly-shaped gland that lies over the heart.

Natural killer cells are cells that resemble T-lymphocytes in many physical properties, although they also show some differences. These cells attack virus-infected cells and tumor cells and kill them. Natural killer cells exist in normal individuals who have not previously encountered the infectious agent or cancer; this is different from the situation for B- and T-lymphocytes, as we shall discuss. Individual natural killer cells are not specific for the cells they attack, but they recognize general characteristics of infected cells or tumor cells. This distinguishes them from B- and T-lymphocytes (as discussed later in this chapter).

The Lymphatic Circulation

Lymphocytes (both B-cells and T-cells) circulate through the blood vessels and also through a second circulatory system, the *lymphatic circulation*. The lymphatic

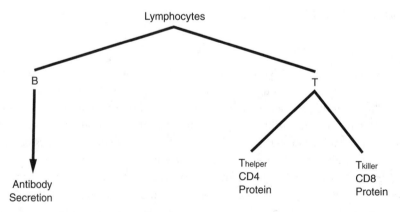

Figure 3-5   Kinds of lymphocytes.

circulation is made up of *lymph channels* in our tissues, which drain lymph fluid from the tissues into structures called *lymph nodes* (Figure 3-6). The lymph nodes contain B-lymphocytes and T-lymphocytes; they can respond to foreign antigens during infections. As an example, suppose a tissue becomes infected with a virus. Pieces of virus or whole virus particles will be transported in the lymph fluid down the lymph channels to the lymph node. In the lymph node, the virus may be recognized as an antigen by B- or T-lymphocytes, which respond by secreting antibodies specific for the virus or by producing T-lymphocytes specific for the virus. These antibodies and lymphocytes drain from the lymph node through another lymph channel, which joins other lymph channels from other parts of the tissue. Ultimately, fluid from lymph nodes all over the body is collected in a series of lymph vessels that empty into a main vessel called the *thoracic duct*, which empties into the bloodstream. As a result, antibodies and lymphocytes that are produced in response to an infection at one site or tissue will be distributed by the bloodstream throughout the body.

During infections, the lymph nodes near the site of the infection frequently become enlarged. This is because the lymphocytes in these lymph nodes are dividing rapidly and producing large amounts of antibodies and cells to fight the infectious agent. For example, you may have noticed swollen glands in your neck if you have had a respiratory infection. The spleen, another organ in the body, has many of the same

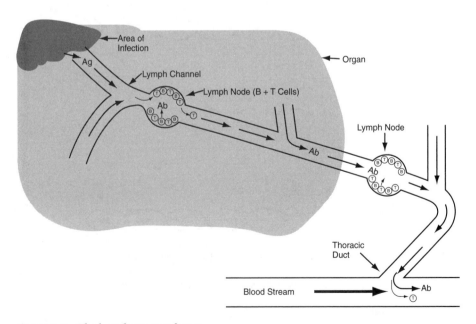

Figure 3-6   The lymphatic circulation.

cells that a lymph node has. These spleen cells carry out functions similar to those of the lymph nodes.

B-Cells and Humoral Immunity: The Generation of Antibodies

Let us now look at how B-lymphocytes respond to a foreign antigen by making antibodies. This part of the immune system is referred to as *humoral immunity* because it results in production of antibodies that circulate in the bloodstream. *Humor* is derived from the Latin word for fluid.

Antibodies

An antibody molecule is made up of four proteins that are bound together: Two of these proteins are identical and are called *heavy chains*; the other two are also identical and are called *light chains*. A protein is a linear chain of building-block molecules called amino acids—much like beads on a string. There are 20 possible amino acids, and the nature of a protein is determined by the particular sequence of the amino acids it contains (Figure 3-7). In the case of antibodies, the two heavy chain proteins are larger than the two light chain proteins. These proteins are held together by chemical bonding into a Y-shaped molecule, as shown in Figure 3-8. Each antibody molecule is specific for one particular antigen, and this specificity is determined by the sequence of

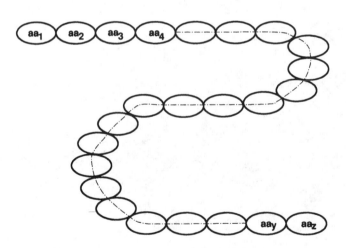

aa1=amino acid number 1 in the protein chain

aa2=amino acid number 2 in the protein chain

etc.

Figure 3-7   Protein structure.

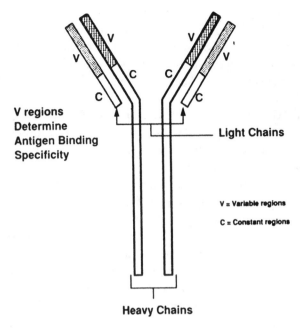

V regions Determine Antigen Binding Specificity

Light Chains

V = Variable regions

C = Constant regions

Heavy Chains

Figure 3-8   Structure of an antibody molecule.

amino acids in the light and heavy chains. If several different antibodies with different specificities are compared, certain regions of the light and heavy chains are very similar for the different antibodies in terms of their amino acids. These regions are referred to as *constant regions*, or C-regions. Other parts of the light and heavy chain proteins are different for each different antibody in terms of the amino acid building blocks. These parts are called the *variable regions*, or V-regions. The protein sequences of the variable regions determine which antigen the antibody binds to. An antibody fits its antigen as a key fits only its own lock. Once an antibody is bound to its proper antigen, the C-region then signals other parts of the immune system to attack—for instance, phagocytosis by a neutrophil or macrophage (see Figure 3-4).

One important feature of the humoral immune system is that *each B-lymphocyte makes only one type of antibody*, with a single specificity for an antigen. Thus, each B-lymphocyte is specific for one antigen.

How Does the Immune System Respond to New Antigens?

During our lives, the number of different infectious agents and antigens that we might encounter is infinite. To protect us from disease, our immune system must be able to respond to each new antigen on demand by making new antibodies that recognize it.

On the other hand, it is impossible for the immune system to anticipate all possible antigens and continually make all possible antibodies that might be required all the time. This would be much too costly in terms of energy and genetic material. To solve this dilemma, the immune system uses two processes: generation of antibody gene diversity by DNA rearrangement and clonal selection.

Generation of Antibody Gene Diversity by DNA Rearrangement

The genetic information for the antibody proteins is contained within DNA in our chromosomes. DNA is a long molecule made up of two strands wound around each other—a "double helix." Each strand is a chain made up of building blocks called *nucleotides*, which contain four possible *bases* (adenine, or A; cytosine, or C; thymine, or T; and guanosine, or G). The exact order of bases in a DNA molecule specifies the order of amino acid building blocks in the corresponding protein, as shown in Figure 3-9. The sequence of DNA bases that specifies one protein is referred to as a gene. Each of our chromosomes contains many thousands of genes along its DNA molecule. We inherit two sets of DNA molecules in the form of chromosomes—one set from our mother and one set from our father. The DNA content of most of the cells in the body is the same—different kinds of cells make different kinds of proteins by selecting which genes will be expressed by way of messenger RNA (mRNA; see Chapter 4) for synthesis into protein. However, antibody-producing B-lymphocytes are an exception, at least as far as the region of the chromosome that specifies antibody proteins is concerned.

It is important to remember that each mature B-lymphocyte produces only one kind of antibody. Thus, each B-lymphocyte makes one kind of heavy chain protein and

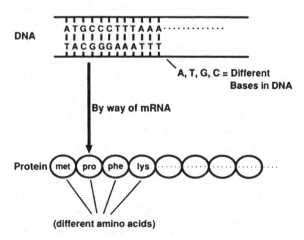

Figure 3-9   How genetic information in DNA is converted into protein.

one kind of light chain protein. All the cells in the body actually contain multiple copies of the genes for variable regions of the heavy and light chain proteins. For the heavy chains, the variable region is actually expressed from three sets of genes called V-genes, D-genes, and J-genes. There are about 200 different V-genes, about 50 different D-genes, and about 10 different J-genes. During development and maturation of a B-lymphocyte, the DNA in the chromosomes surrounding the antibody genes is rearranged (Figure 3-10). As a result of the rearrangement, one V-gene is brought together with one D- and one J-gene, and this combination is next to the gene for the constant region. The intervening V, D, and J DNA sequences are deleted. This VDJ combination is expressed along with the constant region gene to give the heavy chain protein. Light chain protein also results from a similar DNA rearrangement process, except that the variable region is specified by only two sets of multiple genes, V-genes and J-genes.

The DNA rearrangements of the antibody genes (VDJ for heavy chain and VJ for light chains) in any individual developing B-lymphocyte are *randomly selected* from the various possible V-, D-, and J-genes. Thus, the total number of possible VDJ combinations for the heavy chains in a B-lymphocyte is the *product* of the number of V-genes times the number of D-genes times the number of J-genes (200 V-genes × 50 D-genes × 10 J-genes = 100,000 combinations for the variable region). Similarly, the total possible VJ combinations for light chain proteins is the product of the number of light chain V-genes times the number of light chain J-genes. Because each B-lymphocyte produces antibody containing one heavy chain and one light chain, the total number of possible antibodies a B-lymphocyte can make is the product of the possible kinds of heavy chain proteins times the possible kinds of light chain proteins. Thus, the number of possible antibodies that a B-lymphocyte could make is many millions.

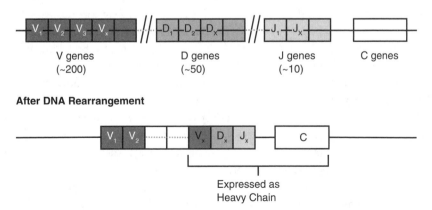

Before DNA Rearrangement

V genes
(~200)

D genes
(~50)

J genes
(~10)

C genes

After DNA Rearrangement

C

Expressed as
Heavy Chain

Figure 3-10   DNA rearrangement for expression of antibodies.

Another process takes place during B-lymphocyte maturation in addition to the DNA rearrangement of the antibody genes. Individual DNA bases in the genes for the variable regions may be changed or added. These changes will further alter the amino acid sequences of the variable regions for the light and heavy chain proteins. Because these changes also occur on a random basis, they *further increase* the number of kinds of variable regions on the antibody proteins. In practice, the number of possible kinds of antibody proteins that can be made is almost limitless.

Clonal Selection

In a normal uninfected individual, many different B-lymphocytes have each carried out the DNA rearrangements of their antibody genes, and more mature every day. Initially, these B-lymphocytes express their specific antibodies on their outer surfaces, but they do not secrete antibody and they do not divide; at this stage they are immature. However, if a particular immature B-lymphocyte recognizes an antigen that binds to its specific antibody (for instance, a protein from an infecting virus), it receives a *signal for activation*. Other B-lymphocytes that are present but that have not bound an antigen do not receive the activation signal. If the B-lymphocyte that has bound an antigen also receives a *second signal* (discussed later), it becomes *fully activated* and mature (Figure 3-11). A fully activated mature B-lymphocyte does two things: It divides rapidly, generating more activated B-cells that make the same antibody, and these activated B-cells all secrete the specific antibody into the extracellular space (for instance, the lymph or blood). The result of this process is the production of large amounts of antibody specific for the antigen.

The Primary Immune Response

The *primary immune response* occurs when the immune system encounters an antigen for the first time, as shown in Figure 3-12. For several days after an antigen is encountered, there are no antibodies for the antigen in the bloodstream. This lag period can last for as little as 10 days or as much as several weeks. During the lag period, B-lymphocytes are being primed with antigen and activated to divide and produce antibody. Eventually, antibodies specific for the antigen begin to appear in the bloodstream and increase until they reach a plateau level. Then, if the antigen is eliminated, the antibody level slowly falls until it returns to an undetectable (or barely detectable) level.

In terms of infectious agents such as viruses and bacteria, the lag period during the primary immune response is very important. During this period, no antibodies against the microorganism are being produced. Thus, the individual is susceptible to continued infection during this period—the immune system will begin to fight most efficiently only after antibodies are produced. This window of vulnerability is particularly critical for viral infections because it is often very difficult to eliminate viral infections once they have become established (see Chapter 4, p. 47).

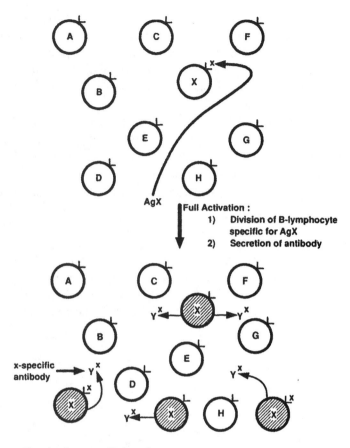

Figure 3-11   Clonal selection of B-lymphocytes.

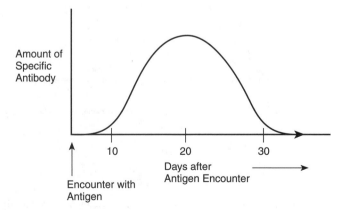

Figure 3-12   The primary immune response.

During the primary immune response, many different B-lymphocytes become primed and activated to produce antibodies. For instance, in the case of a virus infection, B-lymphocytes that make antibodies specific for different virus proteins are activated. Furthermore, different B-lymphocytes may make antibodies for different parts of a single virus protein. All these B-lymphocytes contribute to the mixture of antibodies that makes up the immune response to this virus.

As the primary immune response progresses, the quality of the antibodies also improves. Those antibodies whose variable regions bind most tightly to the antigen become predominant. In addition, the nature of the constant regions of the antibody molecules changes. This change leads to more efficient signaling by the antibodies to other cells of the immune system (such as phagocytes) for attack and destruction of the foreign cell or microorganism.

The Secondary Immune Response

A *secondary immune response* occurs in individuals who have previously raised an immunological reaction against a particular antigen—for instance, someone who has recovered from an infection and then later encounters the same infectious agent. In this case, the levels of specific antibodies rise very rapidly, almost without a lag (Figure 3-13). The levels of specific antibody also fall more slowly than after the primary

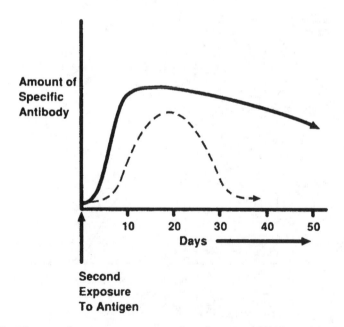

Figure 3-13 The secondary immune response (———) compared with a primary immune response (– – – –).

immune response. In addition, the antibodies are of the high-quality kind, which bind antigen tightly and efficiently signal to other immune cells for attack. Thus, the immune system is said to have immunological memory—the ability to respond rapidly and efficiently to an antigen that has been encountered previously.

Vaccines

The nature of the primary and secondary immune responses and immunological memory have led to development of *vaccines* and *vaccination* for controlling infections. The principle is to preexpose an individual to part of an infectious agent that cannot cause disease (the vaccine) and to induce production of antibodies against that agent. Repeated injections during the initial immunization will induce the production of high-quality antibodies. After the initial immunization, booster injections at regular intervals stimulate the immunological memory and maintain circulating antibodies for the infectious agent. In a vaccinated individual, these antibodies prevent the infectious agent from establishing itself.

Tolerance

Normal tissues in our bodies contain many molecules that could possibly serve as antigens for our own immune systems. It would be very detrimental to our health if our immune systems attacked our own tissues. Indeed, there are immunological disorders, called *autoimmune diseases*, that result from the immunological attack of an individual's own tissue (for instance, rheumatoid arthritis). In normal individuals, the immune system distinguishes between *self* and *non-self*. This is achieved by the development of *tolerance* toward normal tissues. For the most part, this is accomplished by elimination of B- and T-lymphocytes that recognize normal tissues during early development. Because these self-specific lymphocytes are absent, no immunological response toward normal tissue will occur. In addition, other T-lymphocytes provide a second line of defense, should some self-specific lymphocytes avoid elimination. These lymphocytes are called $T_{suppressor}$ lymphocytes, and they prevent B-lymphocytes or T_{helper} lymphocytes specific for self-antigens from maturing. $T_{suppressor}$ lymphocytes have CD8 protein on their surfaces, like T_{killer} lymphocytes.

A Summary of the Humoral Immune System

We summarize the humoral immune system as follows:

1. B-lymphocytes make antibody molecules, and each B-cell makes only one kind of antibody.
2. The immune response is based on:
   a. Generation of many B-lymphocytes with different antibody specificities by DNA rearrangement and mutation within the antibody genes, and

b. Clonal expansion of B-cells that recognize their specific antigen when infection occurs.

3. Antibodies fight infections by:
   a. Direct neutralization of viruses,
   b. Binding to targets and signaling phagocytes or other white blood cells to attack, or
   c. Binding to target cells and signaling for other host defense mechanisms.

T-Cells and Cell-Mediated Immunity

As previously described, T-cells make *T-cell antigen receptors* that resemble antibodies made by B-cells. As with an antibody, the T-cell receptor is composed of four proteins—two large (α) and two small (β) chains. The α and β chains both have constant and variable regions, and the variable regions determine the receptor's specificity toward an antigen. Also like B-lymphocytes, each T-lymphocyte makes only one kind of T-cell antigen receptor. Thus, each T-lymphocyte is specific for a particular antigen. T-lymphocytes do not release their receptors; instead, the receptors are anchored in the cell surface, and the variable regions project outside. As a result, T-lymphocytes will bind to cells expressing antigen by way of their T-cell antigen receptor. T-lymphocytes comprise *cell-mediated immunity* because the cells themselves specifically bind with antigens. This method contrasts with humoral immunity, in which antibodies released from B-lymphocytes carry out the antigen binding.

T_{killer} Lymphocytes

T_{killer} lymphocytes (also called cytotoxic T-lymphocytes) bind cells carrying a foreign antigen and directly kill those foreign cells. Once they have carried out this killing, they release from the target cell, which has been destroyed, and can bind and kill other cells. An example of such an interaction is shown in Figure 3-14. Some examples of cells that T_{killer} cells attack include the following:

1. *Virus-infected cells.* Most cells infected with viruses express some of the viral proteins (or parts of them) on their outer surfaces. These viral proteins can be recognized as foreign antigens and bind T_{killer} lymphocytes. As a result, the virus-infected cells are killed.

2. *Tumor cells.* When cancers develop, they often express abnormal proteins on their outer surfaces. These abnormal proteins can also provoke an immune response by T-lymphocytes, which results in immunological attack on the tumor cells. In fact, the immune system is an important part of our natural defense against cancer. During our lives, probably many cells in our bodies begin to develop into tumors, but the cell-mediated immune system

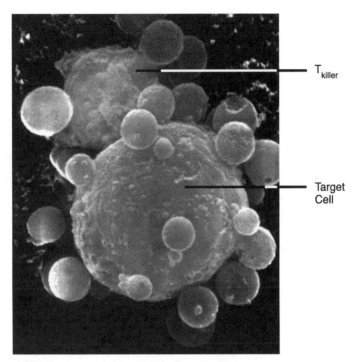

T$_{killer}$

Target
Cell

Figure 3-14   Electron microscope picture of T$_{killer}$ lymphocyte killing a target cell (the largest cell). (© Science Photo Library/Photo Researchers, Inc.)

eliminates them before they can grow very much. This process is called immunological surveillance. It is also important in AIDS because, as we shall see, failure of the immune system can result in development of cancers. In addition to T-lymphocytes, natural killer cells (discussed earlier) are also very important in immunological surveillance.

3. *Tissue rejection.* When tissue from an unrelated individual is introduced into another person, the cell-mediated immune system will generally raise a strong response and kill the transplanted tissue. This reaction is because cell surface proteins, called histocompatibility antigens, generally differ from one individual to another. When tissue with different histocompatibility antigens is transplanted into an individual, a strong cell-mediated immune response against these antigens occurs, and the transplanted tissue is destroyed. For successful organ transplantation, donors and recipients must be carefully matched for histocompatibility antigens to avoid rejection of the donor organ. Even then, the recipients must take immunosuppressive drugs (drugs that suppress the immune system) permanently to avoid rejection of the donated organ.

T_{helper} Lymphocytes

T_{helper} lymphocytes play a central role in both humoral and cell-mediated immunity. In *humoral immunity*, they provide the second signal necessary for a B-lymphocyte that has bound antigen to divide and secrete antibodies (see Figure 3-11). In fact, for a B-lymphocyte that has bound antigen to become fully activated, a T_{helper} lymphocyte with the *same antigenic specificity* must bind the antigen as well, as shown in Figure 3-15. Once the specific T_{helper} lymphocyte is bound to the B-lymphocyte by way of the antigen, it provides growth and maturation signals to the B-cell in the form of growth factors, leading to cell division and antibody production. If a T_{helper} lymphocyte of the same antigen specificity as the B-lymphocyte is absent, the B-lymphocyte will not complete maturation, even if it has bound antigen.

T_{helper} cells also play an important role in *cell-mediated* immunity. When T-lymphocytes (either T_{helper} or T_{killer}) bind antigen, they become activated to divide. This results in increased numbers of specific T-lymphocytes to fight the foreign infectious agent. However, as for many blood cells, T-lymphocytes also need a growth factor to divide. For T-lymphocytes that have bound antigen, the required growth factor is one called *interleukin-2* (IL-2). It turns out that T_{helper} lymphocytes produce and secrete IL-2 when they are activated by antigen binding (Figure 3-16). Thus, the

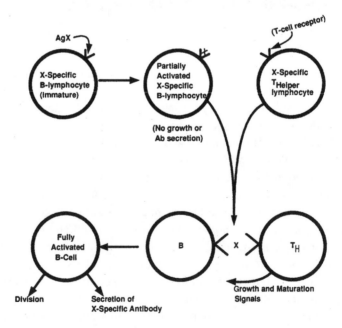

Figure 3-15   The role of T_{helper} cells in B-lymphocyte activation.

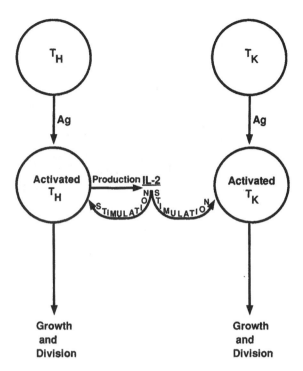

Figure 3-16 T_{helper} cells in cell-mediated immunity.

T_{helper} lymphocytes can stimulate themselves to divide after they bind antigen. On the other hand, most T_{killer} cells do not produce IL-2 even after they bind antigen. They generally rely on IL-2 secreted by neighboring T_{helper} cells to divide. In this case, the neighboring T_{helper} cell that produces the IL- 2 does not have to be specific for the same antigen as the T_{killer} cell it helps. Thus, if T_{helper} lymphocytes are absent, T_{killer} cells cannot divide even if they have bound their specific antigens.

In fact T_{helper} lymphocytes can be subdivided into T_{H1} and T_{H2} cells. T_{H1} cells provide help to T_{killer} lymphocytes (e.g., through secretion of IL-2), while T_{H2} cells provide help to B-lymphocytes by secretion of other growth factors.

In summary, T_{helper} lymphocytes play a central role in both humoral and cell-mediated immunity, as illustrated in Figure 3-17. As we shall see in Chapter 4, the major problem in AIDS is that the causative agent, HIV, specifically infects and kills T_{helper} lymphocytes. This causes a failure of both the humoral immune system and cell-mediated immunity. As a result, immunological protection against infectious agents and cancers is impaired.

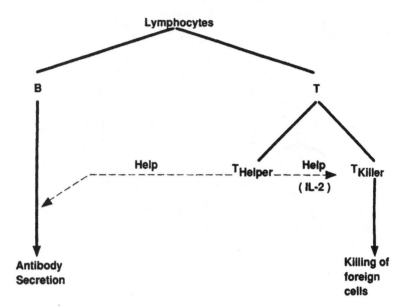

Figure 3-17   The central role of T_{helper} lymphocytes.

T-Lymphocyte Recognition of Antigen-Containing Cells

As we saw earlier in this chapter, when B-lymphocytes recognize their antigens, the antigens are freely circulating in the extracellular space (see Figure 3-11). In contrast, when T-cells recognize their specific antigen, they recognize a digested form of the antigen that is held on the surface of a cell. For T_{killer} lymphocytes, these cells recognize cells expressing a "foreign" antigen. For instance, a virus-infected cell is expressing viral proteins. Portions of the viral proteins (X) are digested into small fragments ["(X)"] that are transported to the surface of the cell where the fragments are held in a protein complex called *major histocompatibility complex class I* (MHC-I) (Figure 3-18). If a T_{killer} lymphocyte specific for antigen X encounters the infected cell, then the T-cell receptor on the surface of the T_{killer} lymphocyte will specifically bind the MHC-I–(X) complex. This specific binding will lead to killing of the infected cell. Most cells in the body express MHC-I, so they can potentially be killed by T_{killer} lymphocytes if they express an antigen.

When T_{helper} lymphocytes encounter antigen-containing cells, a similar but distinct process takes place. The antigen-containing cells that can interact with T_{helper} lymphocytes are called *antigen-presenting cells* (Figure 3-19). A common feature of antigen-presenting cells is that they express *major histocompatibility complex class II* (MHC-II). Another feature of antigen-presenting cells is that they can take in antigens

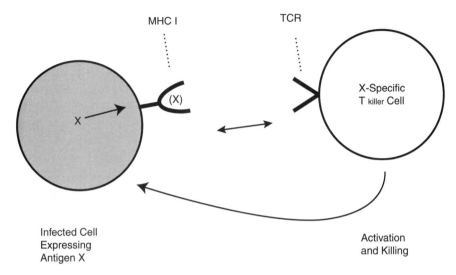

MHC I   TCR

(X)

X

X-Specific
T killer Cell

Infected Cell
Expressing
Antigen X

Activation
and Killing

Figure 3-18 T_{killer} recognition of antigen-expressing cells.

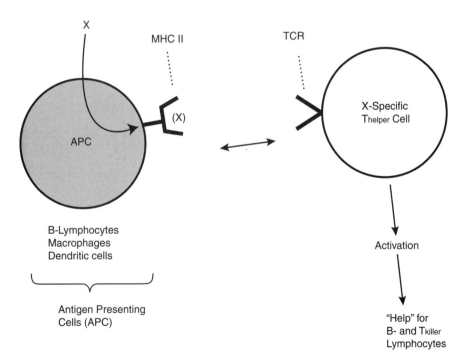

X

MHC II   TCR

(X)

APC

X-Specific
Thelper Cell

B-Lymphocytes
Macrophages
Dendritic cells

Antigen Presenting
Cells (APC)

Activation

"Help" for
B- and Tkiller
Lymphocytes

Figure 3-19 TCR, T_{helper} recognition, of antigen-presenting cells. See Figures 3-15 and 3-16 for the mechanisms of help for B-lymphocytes and T_{Killer} lymphocytes.

from the extracellular space, digest the antigens ["(X)"], and transport them to the cell surface where they are held in an MHC-II complex. If a T_{helper} lymphocyte specific for antigen X encounters an antigen-presenting cell displaying digested antigen X, then specific binding between the T-cell receptor on the T_{helper} lymphocyte and the MHC-II–(X) complex will occur. This binding leads to activation of the T_{helper} lymphocyte and provision of "help" to B-lymphocytes and T-lymphocytes, as diagrammed in Figure 3-17. Depending on whether the T_{helper} cell is a T_{Hh1} or T_{Hh2} cell, help will be provided to T_{killer} or B-lymphocytes.

There are three kinds of antigen-presenting cells in the body: B-lymphocytes, macrophages, and *dendritic* cells. The interaction between B-lymphocytes and T_{helper} lymphocytes in production of antibody was previously discussed and diagrammed in Figure 3-15. If a macrophage engulfs a virus-infected cell, it can "present" digested virus proteins on its surface where specific binding of antigen-specific T_{helper} lymphocytes will lead to activation of the T_{helper} cells and provision of "help" to surrounding B-lymphocytes and T-lymphocytes. Dendritic cells are interesting because they exist in the skin and at the surface of internal organs such as the large intestines. These cells have long processes that extend through tissues and even to the mucosal surfaces of the gut. Dendritic cells take up antigens, digest them, and present them at the surface in MHC-II complexes. Another important feature of dendritic cells is that once they have bound antigen, they can migrate through tissue or blood to nearby lymph nodes, where there are many T_{helper} lymphocytes. Thus, they carry antigens from the skin or mucosal surfaces to the lymph nodes for efficient induction of immune responses.

---

http://biology.jbpub.com/fan/aids/6e/

Connect to this book's website: http://biology.jbpub.com/fan/aids/6e/. The site features summaries of the main points from each chapter, links to important AIDS-related websites, and short-answer-style review questions for each chapter.

CHAPTER 4

Virology and HIV

In this chapter, we first look at viruses in general, then retroviruses, and then HIV—the virus that causes AIDS—in particular. We will also see how the HIV antibody test (used in screening for HIV infection) works and what it tells us. We will then consider the basis of action of the drug azidothymidine (AZT) and of protease inhibitors, which are currently used as antiviral treatments for HIV infection.

A General Introduction to Viruses

Let us first consider viruses in a general sense. There are many different kinds of viruses, and many of them cause disease. Individual viruses may differ in their exact composition and mechanisms for growth, but all viruses have some common properties.

What Are Viruses?

Viruses are among the simplest life forms. Here are some of the common features of viruses:

1. *Viruses are obligate intracellular parasites.* This means that viruses cannot replicate and make more of themselves outside cells. In fact, a pure preparation of virus particles will not grow. In the case of humans, this means that viruses must replicate in some tissue or cell type in our bodies.

2. *Virus particles consist of the following components* (Figure 4-1):

   a. *Genetic material.* Viruses carry genetic material in the form of nucleic acids. For some viruses, the nucleic acid is DNA, the same as the genetic material of the cells in our bodies. For other viruses, the nucleic acid is RNA, which is related chemically to DNA (discussed later). *The genetic material of a virus specifies virus proteins.* These virus proteins may be *structural proteins* that make up the virus particles, *enzymes* that help carry out biochemical processes necessary for virus growth, or *regulatory proteins.* Some viral regulatory proteins are used by the virus to select expression of particular virus genes at different times or under different conditions. Other viral regulatory proteins may be used by the virus to help it take over the cell and convert it into an efficient factory for producing the virus.

   b. *A system for protecting the genetic material and introducing it into a cell.* Viruses must protect their genetic material when they leave one cell

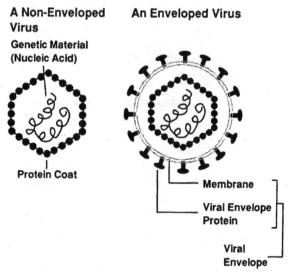

Figure 4-1   Structure of a typical virus.

and move to another—either within tissues of an infected individual or from an infected individual to an uninfected one. Naked DNA or RNA is quite fragile and vulnerable to attack by numerous agents. Thus, viruses carry genes that direct production of a *protein coat* that surrounds the genetic material. In addition, some (but not all) viruses direct synthesis of a *viral envelope* that surrounds the virus's genetic information and protein coat. Viral envelopes resemble the membranes that make up the outer surfaces of our cells. These envelopes contain proteins that are virus specified. For viruses that contain envelopes, the envelope proteins are very important for the initial phases of infection because they are exposed on the outside of the virus particle.

3. *Viruses are dependent on cells for the following*:
   a. *Energy metabolism.* Energy is required for most biochemical processes to take place. In the case of viruses, such processes include those responsible for production of the virus's proteins and genetic material. However, viruses themselves do not carry the machinery necessary for generating energy. Instead, they rely on the machinery of the cells they infect.
   b. *Protein synthesis.* Proteins are synthesized in cells by a complex system of molecules and subcellular particles, using instructions from the genetic material. Again, viruses carry the genetic instructions but do not carry the machinery for synthesis of proteins. They depend on the cell protein synthesis machinery.
   c. *Nucleic acid synthesis.* Many viruses may also depend on the cell machinery for synthesis of virus-specific nucleic acids. These nucleic acids may be used for expression of viral proteins (mRNA, discussed later), or they may be the virus's genetic information itself.

How Does a Virus Infect a Host?

For a virus to infect an individual, it must come into contact with a susceptible cell. It is important to remember that most of the human body is covered with skin, which protects us from infection. Skin is quite tough, and the outer layers of skin cells are actually dead. Thus, most viruses cannot infect and grow in cells of the outer layers of the skin. The following are some of the important routes that viruses use to enter the body (Figure 4-2):

The respiratory tract: Viruses can be carried into the respiratory tract through the air we breathe. Once they are brought into the body by this route, they can infect cells in any part of the respiratory tract, including the nose, windpipe, bronchial tubes, and lungs. Examples of viruses that infect the respiratory tract are influenza and the common cold.

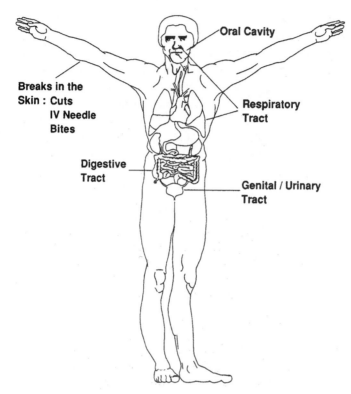

Figure 4-2   Routes of entry for viruses. IV = intravenous.

The oral cavity and digestive tract: If viruses are taken in with food or water, they can potentially infect cells of the mouth and other parts of the digestive system, including the large and small intestines. One form of liver inflammation, hepatitis (infectious or type A hepatitis), is an example of this category, as are various diarrheas.

The anal/genital tract: During sexual intercourse, it is possible to introduce viruses into the female or male anal/genital tract from an infected partner. Such infections are classified as venereal diseases. If sexual intercourse involves anal penetration, it is possible to introduce viruses into the anus, rectum, and lower intestines by this route as well. Genital herpes virus is an example of an infection of the genital tract. As we shall see, genital tract infection is an important route for HIV and AIDS.

Breaks in the skin: If the protective layer of skin is broken by a cut or scratch, then viruses may be able to enter directly into tissues or the bloodstream. Bites from animals or insects also fall into this category. For example, rabies is spread by bites from infected animals such as dogs or squirrels, and yellow fever is spread by bites from infected mosquitoes. Transfusions and injection drug use are other examples of infection through breaks in the skin. In these cases, viruses that contaminate blood or blood products can be introduced into individuals during transfusions with blood or blood products or during injection drug use involving shared needles. For example, another form of hepatitis, hepatitis B, can be spread by injection drug use (as well as by sexual contact). Injection drug use (and, originally, transfusions) is another important route of infection for HIV.

It is important to remember that any individual type of virus will use some but not all of these routes of infection. A key to controlling viral infections is understanding the particular routes of spread the virus of interest uses. We shall see how this is determined in Chapter 6. After a virus has entered an individual and established infection at a *primary site*, the infection can spread to *secondary sites* in the body as well. Disease symptoms may result from infection at the primary site, the secondary sites, or both.

A Typical Virus Infection Cycle

Let us look at what happens if a purified virus preparation is used to infect some susceptible cells in the laboratory. A typical result is shown in Figure 4-3. If the amount of infectious virus is measured over a period of time, it is seen to fall after an initial lag period, remain low for a period of time, and then rise to even higher levels.

The period during which the amount of infectious virus is low is referred to as the *eclipse period*. The virus infection cycle can be divided into several events:

1. *Adsorption (binding) of the virus to the cell.* When a virus infects, it must first bind to the cell. This binding is a very specific interaction between the virus particle and some protein (or other molecule) on the cell surface. This protein is referred to as the virus *receptor*. At first, it might seem strange that cells have receptors for viruses, because this would seem to be disadvantageous to the uninfected host. However, viruses have evolved so that they are able to bind to a protein that is normally present on the uninfected cell. The distribution of the receptor protein among different cells in the body influences the kinds of cells that the virus can infect. We will see that this is an important consideration for HIV infection and development of AIDS.

2. *Penetration of the virus into the cell and uncoating of the viral genetic material.* Once the virus particles have bound to the surface of the cell by attaching to a receptor protein, they are brought into the cell. This penetration process is an active one that requires expenditure of energy by the cell. Once the virus particle has been taken into the cell, its protective protein coat is removed, exposing the viral genetic material. The genetic material is now ready to be expressed.

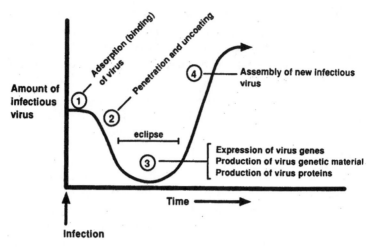

Figure 4-3   A typical virus infection cycle within a cell.

This uncoating of the virus accounts for the drop in infectious detected virus because the uncoated virus cannot withstand the assay conditions.

3. *Expression of the viral genetic material.* This occurs during the eclipse period, when the amount of infectious virus in the culture appears low. Several events take place during the eclipse phase:

   a. *Organization of the infected cell for virus expression.* The cell machinery may be altered to favor efficient expression of virus genes. This often occurs at the expense of the cell's own metabolic processes and may ultimately lead to death of the infected cell.

   b. *Replication of the viral genetic material.* The virus programs the machinery necessary to generate more copies of its own genetic material. In some cases, it may rely on machinery from the uninfected cell, but in other cases, the virus may specify proteins that are necessary for the process.

   c. *Synthesis of proteins for virus particles.* Proteins that make up the virus coat, as well as those in the viral envelope, are synthesized from instructions in the viral genetic information. Once these proteins are synthesized, all the components necessary for formation of a new virus particle are present within the infected cell.

4. *Assembly of virus particles and release from the cell.* Virus particles are assembled in the infected cell from the new genetic material and viral proteins. As this occurs, the amount of infectious virus in the culture increases and surpasses that at the start of the infection. Typically, an infected cell releases hundreds or thousands of new virus particles, which can spread to infect other cells.

Depending on the virus, there are different fates for an infected cell. For many viruses, the infected cell is killed (or lysed) at the end of the infection. These viruses are called *lytic*. Other viruses do not kill the infected cell, but they establish a persistent or carrier state in which the cell survives and continually produces virus particles. These viruses are called *nonlytic*. Some viruses can also establish a state called *latency* in cells. In these situations, the virus's genetic material remains hidden in the cell, but no virus is produced. At a later time, the latent virus can become *reactivated*, and the cell will begin to produce infectious virus particles again, as in the case of cold sores caused by a herpesvirus. As we shall see, all these fates play an important role in HIV infection and the development of AIDS.

How Do We Treat Viral Infections?

When viral infections become established, they are very difficult to treat. This contrasts with the wide variety of antibiotics available to treat infections by other microorganisms, such as bacteria and fungi. Antibiotics take advantage of the fact that there are differences in some of the biochemical machinery of these very simple microorganisms compared with highly developed organisms, such as humans. These antibiotics specifically inhibit processes carried out by the bacteria or fungi, but they do not affect similar processes in higher organisms. For instance, the antibiotic streptomycin inhibits the intracellular machinery used to make proteins in bacteria but not in humans. Unfortunately, because viruses rely on the cell to carry out most of their metabolic processes, it is difficult to find drugs similar to classic antibiotics that will block virus growth without killing the infected cell. However, in some cases, compounds that specifically inhibit a viral process have been identified. These compounds are called *antivirals*, and they hold the key for future treatment of viral infections. As we shall see, quite a number of antivirals for HIV have been developed. At the present time, the basic treatment for most virus infections is to manage the symptoms and wait for the infection to run its course. Management of symptoms can include treatment to reduce fevers (for instance, aspirin), classic antibiotics (to prevent secondary infections by bacteria in a weakened individual), and bed rest.

Because treatment of viral infections is difficult, the best approach to managing viral disease is to prevent the initial infection. One powerful method is public health and sanitation methods to intervene in the epidemiological cycle of the virus, as described in Chapter 2. Another important approach is the use of viral vaccines, as described in Chapter 3. If immunity to a virus can be induced by the vaccine before a person encounters the virus, then it will not be able to establish a foothold. Some of the best-known virus vaccines are the smallpox vaccine developed by Edward Jenner (the first vaccine), the rabies vaccine developed by Louis Pasteur, and the polio vaccines developed by Jonas Salk and Albert Sabin.

The Life Cycle of a Retrovirus

HIV belongs to a class of viruses called *retroviruses*. Let us examine the life cycle of a typical retrovirus.

The structure of a retrovirus is shown in Figure 4-4. The genetic information of a retrovirus is RNA. This RNA is covered with a viral protein coat; together, the viral RNA and protein coat make up a core particle. The core particle also contains several virus-specified enzymes. The core particle is surrounded by a viral envelope, which contains membrane lipids and viral envelope protein.

All retroviruses have three genes (Figure 4-4). These genes code for the following:

1. *Coat proteins that make up the inner virus (core) particle.* The virus gene that specifies these proteins is called the *gag gene.* For HIV, there are three *gag* proteins, p17 (or MA), p24 (or CA), and p10 (or NC).
2. *The enzyme reverse transcriptase, as well as other enzymes used in virus replication.* The gene that codes these enzymes is the *pol* gene. The other viral enzymes specified by the *pol* gene are *protease* and *integrase.* Protease

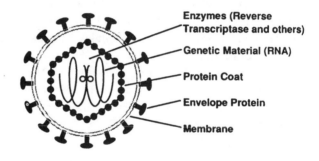

The RNA Genetic Material

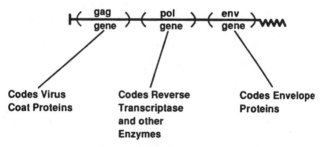

Figure 4-4   The structure of a retrovirus and its RNA genetic material.

is involved in maturation of viral proteins as the virus particles bud from the cell, and integrase is responsible for integration of the viral DNA into the cell's chromosomal DNA.

3. *The proteins of the viral envelope.* The gene that codes for these proteins is the *env* gene. A protein coded by the *env* gene is responsible for binding the virus to the cell receptor. For HIV, there are two *env proteins*, gp120 and gp41.

It is important to discuss the *central dogma for genetic information flow* in cells. The central dogma states that genetic information flows in the following direction:

DNA → RNA → Protein

That is, the genetic information is carried in DNA as a sequence of nucleotide bases (see Figure 3-9). In higher organisms, the DNA is organized into chromosomes that are located in the *nucleus* of the cell. When a gene is expressed, the information from the DNA base sequence is copied or transferred (transcribed) to a related molecule called *RNA* using the DNA molecule as a pattern. The RNA (which is called *messenger RNA*, or mRNA) then moves from the cell nucleus to the *cytoplasm*. Once in the cytoplasm, the mRNA is used as a blueprint for the formation of proteins (*translation*). The proteins then carry out most of the important functions for the cell.

The life cycle of a retrovirus is shown in Figure 4-5. The retrovirus first binds to the surface of an uninfected cell by recognizing a cell receptor. After binding, the virus particle is brought into the cytoplasm of the cell. During this process, the viral envelope is removed, leaving the core particle. Once this happens, a unique virus-specified enzyme called *reverse transcriptase* is activated. This enzyme reads the viral RNA and makes viral DNA. The host cell lacks such an enzyme. The viral DNA then moves to the nucleus of the cell, where it is incorporated (or *integrated*) into the host cell's DNA in the chromosomes. Viral DNA integration is mediated by the viral integrase protein. Once this viral DNA is integrated into the chromosome, it resembles any other cell gene. As a result, the normal cell machinery reads the integrated viral DNA to make more copies of viral RNA. This viral RNA is then used for two purposes: (1) Some of the viral RNA moves to the cytoplasm and functions as viral mRNA to program the formation of viral proteins, and (2) the rest of the viral RNA becomes genetic material for new virus particles by moving to the cytoplasm and combining with viral proteins. These virus particles are formed at the cell surface and leave the cell by a process called *budding*. When virus particles initially bud from the cell, they are immature. This is because the viral proteins have not assumed their final form. The viral enzyme protease is responsible for conversion of immature virus particles into mature ones.

The retrovirus life cycle has several important characteristics. First, most retroviruses do not kill the cells they infect. Second, because viruses integrate their

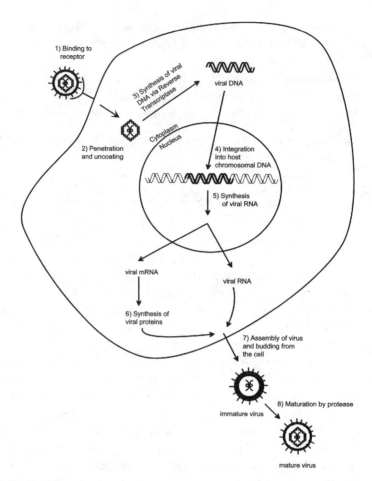

Figure 4-5 The life cycle of a retrovirus.

DNA into host chromosomes, they can establish a stable carrier state within the infected cell. As a result, once cells are infected with most retroviruses, they continually produce virus without dying. For some retroviruses, a latent state may also be established in which the retroviral DNA is integrated into the host chromosomes, but it does not program formation of new virus particles. However, at a later time (sometimes years later), the latent viral DNA may become activated by some means, and virus will be produced. This latency process is important in HIV infection and AIDS.

The viral enzyme *reverse transcriptase* carries out an unusual process in converting the viral RNA genetic information into DNA. This process is the reverse of genetic information flow according to the central dogma of molecular biology, which is why the enzyme is called reverse transcriptase. This is also where retroviruses get their name—*retro* is from the Latin word for reverse.

The AIDS Virus: HIV

As discussed in Chapter 1, the virus that causes AIDS is human immunodeficiency virus (HIV) (Figure 4-6). Other names that have been used previously for HIV include

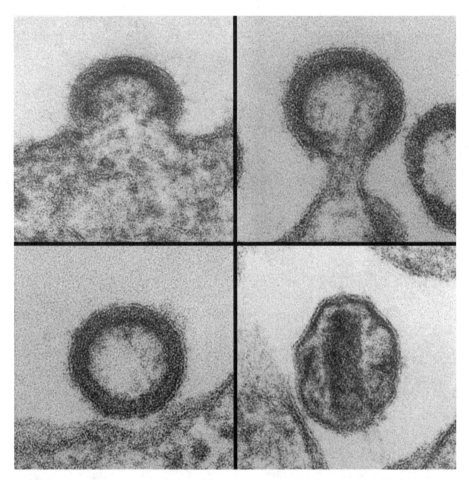

Figure 4-6   An electron microscope picture of an HIV-infected cell. The cytoplasm of the cell is on the bottom, and the exterior of the cell is on the top. Budding HIV particles are shown in the upper panels, and released particles are shown in the lower panels. An immature particle is shown on the lower left, and a mature particle is shown on the lower right. The cores of mature HIV particles have a conical shape. (Courtesy of Matthew A. Gonda, Ph.D., Chief Executive Officer, International Medical Innovations, Inc.)

HTLV-III, LAV, and ARV. HIV belongs to a subgroup of retroviruses called *lentiviruses* (meaning *slow viruses*, because they often cause disease extremely slowly); other lentiviruses have been found in such diverse species as cats, sheep, goats, horses, and monkeys. The virus responsible for the great majority of AIDS cases in the United States, Europe, and Africa is called HIV-1. A second virus related to HIV-1 has been isolated in Africa: HIV-2 (see p. 66 and Chapter 6). HIV-2 also causes AIDS. In this book, we will refer to the AIDS virus simply as HIV, and this will almost always mean HIV-1.

Features of HIV

Several features about the structure and replication of HIV are important (Figure 4-7a).

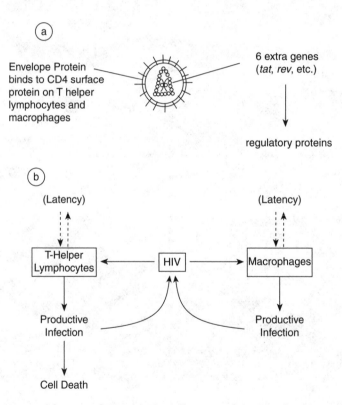

Figure 4-7   Unusual features of HIV. (a) The cell receptor for HIV is the CD4 surface protein, and HIV carries extra genes besides *gag*, *pol*, and *env*. (b) Infection of HIV into T_{helper} lymphocytes results in cell death; infection into both macrophages and T_{helper} lymphocytes can result in latent infection.

The Nature of the HIV Receptor

The cell receptor that HIV binds to is the *CD4 surface protein*. As described in Chapter 3, CD4 protein is present on T_{helper} lymphocytes. In fact, this is the predominant cell type that has CD4 protein. In addition, macrophages and dendritic cells also have CD4 protein. Most other cells in the body do not contain CD4 protein. As a result, the main cells that HIV can infect are T_{helper} lymphocytes, macrophages, and dendritic cells. The HIV envelope protein responsible for virus binding to CD4 protein is called gp120.

Additional Genes

As all retroviruses do, HIV contains the three genes for coat proteins, reverse transcriptase, and envelope proteins (*gag*, *pol*, and *env*). In addition, HIV contains genes that specify six additional proteins (Table 4-1). These are regulatory proteins that give HIV finer levels of control and a more versatile life cycle. Two of the best known of these genes are *tat*, which specifies an up-regulator or amplifier of viral gene expression in the infected cell, and *rev*, which specifies a protein that shifts the balance from production of viral regulatory proteins to proteins that make up virus particles. The other HIV genes specify proteins called *nef*, *vpu*, *vif*, and *vpr*. They facilitate the infection process in different ways. Their roles in HIV infection are summarized in Table 4-1.

Table 4-1	The HIV Genes	
Gene	Protein(s) Specified	Function
gag	MA,CA,NC, p6	Coat proteins of the virus core
pol	Reverse transcriptase	Synthesis of viral DNA
	Integrase	Integration of viral DNA into host cell DNA
	Protease	Cutting viral polyproteins during maturation
env	gp120 (SU)	Binding to the CD4 receptor and co-receptors
	gp41 (TM)	Entry of the viral core after binding
tat	Tat	Positive amplifier of HIV RNA expression
rev	Rev	Shifts expression from regulatory proteins to proteins that make up viral particles
nef	Nef	Necessary for efficient infection
vpr	Vpr	Necessary for efficient infection
vif	Vif	Counteracts a host defense against viruses
vpu	Vpu	Enhances release of new virus particles

Killing of T$_{helper}$ Lymphocytes

In contrast to most retroviral infections, *productive infection of T$_{helper}$ lymphocytes with HIV results in cell death* (Figure 4-7b). When T$_{helper}$ lymphocytes are infected by HIV, productive infection usually results, which leads to production of new HIV virus particles, but the infected cells are also killed. Considering the pivotal role that T$_{helper}$ lymphocytes play in both humoral and cell-mediated immunity (see Chapter 3, p. 36), it is possible to understand how infection with HIV can ultimately lead to collapse of the immune system. In addition, some infected T$_{helper}$ lymphocytes will establish a latent state in which virus is not produced and the cells are not killed. Later, the latently infected cells can be reactivated to produce infectious HIV, at which time they are also killed. The latently infected T$_{helper}$ lymphocytes are a major reservoir of infection in an HIV-infected individual.

To make matters worse, in HIV-infected people, *uninfected T$_{helper}$ lymphocytes are also killed*. The exact mechanism by which this happens is not understood, but this effect contributes to immune system failure in AIDS patients.

Nonlytic Infection of Macrophages

When HIV infects macrophages, it follows a course that is typical of other retroviruses in that productively infected macrophages produce virus and are not killed (see Figure 4-7b). Infected macrophages can also establish a latent state of HIV infection. Macrophage infection is important for the persistence of infection in HIV-infected individuals, both because the infected cells are not killed and because infected macrophages can establish latency.

Dendritic Cells and HIV

Dendritic cells also have CD4 on their surface, so they can be infected by HIV. Like macrophages, they are not killed by HIV infection. In addition, dendritic cells express a protein on their surface called DC-SIGN that can directly bind HIV particles and hold them at the surface without the virus particle entering the dendritic cell. When the dendritic cells migrate from the skin (or surfaces of other organs) to the lymph nodes, they can carry HIV virus particles with them and deliver them to T$_{helper}$ lymphocytes, where infection takes place. Thus, HIV can gain access to T$_{helper}$ cells in the lymph nodes through dendritic cells.

Co-receptors for HIV

As described earlier, when HIV infects a cell, it must bind to the CD4 protein. However, binding to the CD4 protein alone will not result in entry of the HIV virus particle into a cell. The cell must have an additional protein on its cell surface for virus entry to take place—a co-receptor. As shown in Figure 4-8, a cell must express both CD4 protein and a co-receptor to be infected. The *co-receptors for HIV* turn out to be

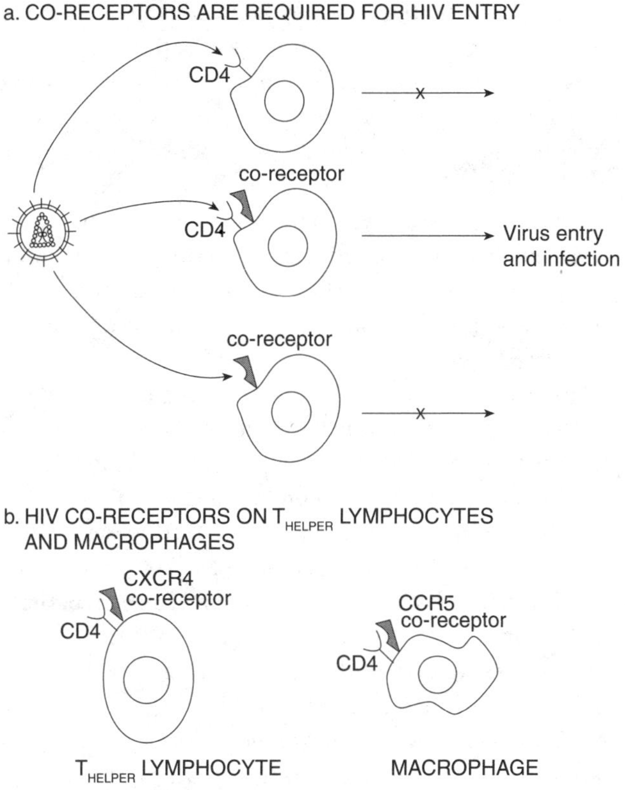

a. CO-RECEPTORS ARE REQUIRED FOR HIV ENTRY

b. HIV CO-RECEPTORS ON T$_{HELPER}$ LYMPHOCYTES
AND MACROPHAGES

Figure 4-8   Co-receptors for HIV infection. (a) In addition to the CD4 protein, cells must also have a co-receptor for HIV infection. (b) The co-receptor on T$_{helper}$ lymphocytes is the CXCR4, and the co-receptor on macrophages is CCR5.

proteins that normally bind particular cell growth factors. As discussed in Chapter 3, specific growth factors are required for the growth of different blood cells (e.g., IL-2 for T$_{helper}$ lymphocytes); these growth factors signal for cell growth by binding to receptor proteins on the cell surface. The HIV co-receptor on T$_{helper}$ lymphocytes is a growth factor receptor called *CXCR4*, and the co-receptor on macrophages is a receptor called *CCR5*.

The Effects of HIV Infection in Individuals

Let us now consider the results of HIV infection at the level of infected people. The routes of HIV infection are covered in Chapters 6 and 7, so here we will start at the time a person becomes infected. There is a detailed description of AIDS as a clinical disease in Chapter 5, but an overview is useful at this point.

The progression of events after HIV infection is shown in Figure 4-9. After HIV infection, there is often (but not always) an acute infection syndrome, typified by a mild flu-like illness or swollen glands, which goes away after a few weeks. Many HIV-infected people do not associate the symptoms of the acute infection syndrome with HIV infection. After the initial infection period, most individuals then remain free of any clinical symptoms for variable lengths of time, typically many years. Individuals who are HIV infected but who do not show any signs of disease are referred to as *asymptomatic.* During the asymptomatic period, individuals generally produce antibodies to HIV. Unfortunately, these antibodies are not sufficient to prevent continued HIV infection as the disease progresses. However, they provide a useful means for diagnosing HIV infection, as we shall see later.

As time passes, many HIV-infected individuals begin to experience symptoms of HIV infection. Some initial symptoms include persistent enlarged lymph glands (*lymphadenopathy syndrome*) and fevers or night sweats. As the disease worsens, a continuum of progressively more serious conditions develops as the immune system weakens, ultimately resulting in full-blown AIDS. During the early periods of the AIDS epidemic, doctors also used a classification called *ARC,* or *AIDS-related complex.* Individuals were classified as having ARC if they showed fewer of the characteristic opportunistic infections or cancers (see next paragraph) than patients with full-blown AIDS. The term ARC is not used today. The progression from asymptomatic infection to AIDS is accompanied by a progressive depletion of T_{helper} lymphocytes by HIV infection. Ultimately, there is a profound lack of T_{helper} lymphocytes, which results in the failure of both humoral and cell-mediated immunity. In individuals

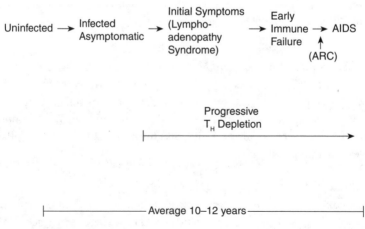

Figure 4-9   Consequences of HIV infection in the absence of antiviral therapy.

with normal immune systems, the T_{helper} counts are typically over 1,000 per cubic millimeter of blood, whereas in patients with full-blown AIDS, these may be well under 100 per cubic millimeter.

The clinical manifestations of AIDS are covered in detail in Chapter 5 but are summarized briefly here:

Opportunistic infections: These are infections by microorganisms that normally do not cause problems in healthy individuals. However, in individuals with weakened immune systems, these microorganisms can take hold and cause devastating infections. One frequent opportunistic infection is *Pneumocystis pneumonia*, caused by a fungus microbe.

Cancers: Cell-mediated immunity also plays an important role in defense against development of cancers (immune surveillance; see Chapter 3, p. 34). HIV-infected individuals develop several cancers with very high frequency. One example of an AIDS-related cancer is *Kaposi's sarcoma*.

Weight loss: Many AIDS patients suffer from profound weight loss or wasting. The mechanism for this is not yet understood.

Mental impairment: HIV can also establish infection in the nervous system. This can result in muscle spasms or tics. More serious is infection of the central nervous system, which can result in *AIDS-related dementia*, in which individuals lose the ability to reason.

Individual AIDS patients may suffer from one or more of these manifestations. Indeed, they may experience recurrent bouts of different opportunistic infections or cancers.

Because the major problem in AIDS is a loss of T_{helper} lymphocyte function, monitoring the numbers of T_{helper} lymphocytes is important in HIV-infected individuals. Doctors can perform a test for these cells, and the results are reported in terms of T_{helper} (or T4 or CD4) lymphocyte numbers.

The likelihood that an HIV-infected individual will develop full-blown AIDS is discussed in more detail in Chapter 6. Without treatment, more than 90% of HIV-infected individuals will develop AIDS with an average time to disease of 10 to 12 years.

One of the features of HIV infection is that until the disease has progressed quite far, relatively little infectious HIV is apparent in the blood of an infected individual. However there is extensive infection of T-lymphocytes in lymph nodes, although the infected cells and the infectious virus are not released into the bloodstream until late in the disease. Sensitive tests can now detect the low levels of virus in the blood of asymptomatic HIV-infected individuals.

The HIV Antibody Test

Within a year of the isolation of HIV as the causative agent of AIDS, a test was developed that determines if an individual has been exposed to HIV. The procedure is to test whether an individual has antibodies to HIV virus proteins. These antibodies appear in those who have been previously infected with HIV and have made antibodies against the virus (see Chapter 3, p. 30).

The most common HIV antibody test is called an *enzyme-linked immunoabsorbent assay (ELISA)* test, shown in Figure 4-10. In an ELISA test, virus protein is first attached to a small laboratory dish. A serum sample is prepared from the blood of the individual to be tested, and it is placed in the dish containing bound HIV viral proteins. If HIV-specific antibodies are present in the serum, they will become tightly bound to the dish by way of the HIV proteins. The serum is then removed, and the dish is washed. During this procedure, only antibodies specific for HIV will be retained. The dish is then reacted with a stain that will detect any human antibodies. Thus, dishes that were exposed to

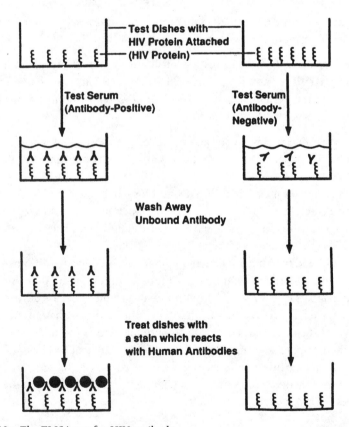

Figure 4-10   The ELISA test for HIV antibody.

serum containing HIV-specific antibodies will be stained, but dishes from antibody-negative serum samples will be unstained. This procedure has been automated, so that many blood samples can be tested at once (Figure 4-11). The current ELISA tests are better than 99.9% accurate. That is, fewer than 0.1% (one in a thousand) of HIV-negative individuals incorrectly score as positive by the ELISA test. Likewise, fewer than 0.1% of HIV antibody-positive serum samples are missed by the test.

Improvements and refinements to HIV ELISA tests are continually being made. For instance, recent improvements include ELISA tests on saliva instead of blood, which is possible because saliva from HIV-infected individuals contains HIV-specific antibodies. Using saliva makes for easier testing because blood samples do not have to be drawn. In addition, recently developed antibody tests can give results in less than 30

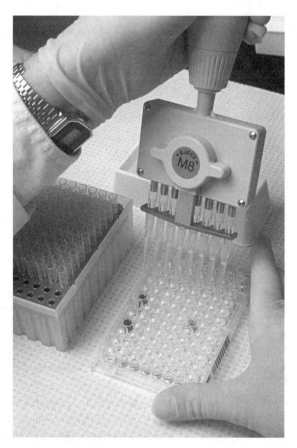

Figure 4-11   A technician uses a multi-pipette to perform an ELISA test for antibodies to the HIV virus. The wells with darker solutions indicate the presence of HIV antibodies. (© Robert Holmgren/Peter Arnold, Inc.)

minutes. Thus, whereas previous antibody tests required the person to make two visits to a clinic—one for the blood sample to be drawn and one to receive the test results—it is now possible to perform the test and give results in one clinic visit.

Potential Problems with the HIV Antibody Test

Although this test has been extremely important in furthering our knowledge of how the virus spreads and causes disease and in identifying HIV-infected people, it has several potential problems.

False Positives

When individuals who are not infected with HIV test positive for HIV antibodies, the results are referred to as false positives. Clearly, this can be extremely frightening. With current ELISA tests, the frequency of false positives is less than 1 in 1,000 (0.1%) uninfected individuals. False positives are a particular problem if populations with low frequencies of HIV infection are tested. In these cases, a high proportion of the individuals who score positive could be false positives. This risk is one of the arguments (besides cost) against routine HIV antibody screening of the general U.S. population, where the current prevalence of infection is less than 1%; many of the individuals identified as antibody positive in such a mass screening could actually be false positives.

To reduce the false positives in the ELISA test, a second, more specific test for HIV antibodies is also used: the *Western blot test*. This technique has a lower incidence of false positives than the ELISA assay. In practice, serum samples that score antibody positive by the ELISA test are retested by the Western blot procedure. Serum samples are considered positive if they are found to contain HIV-specific antibodies by both tests. Other more sensitive tests for HIV infection exist (and will be discussed), but since they are labor intensive, they are not used in screening for HIV infection.

False Negatives

A more important problem is individuals who are infected with HIV but do not score positive in the HIV antibody test. Such individuals fall into two categories:

1. *Recently infected individuals.* As was discussed in Chapter 3, the immune system has a lag period between initial exposure to an antigen and the production of antibodies. In the case of HIV infection, this lag can range up to 6 months or longer. Thus, individuals who have been recently infected with HIV will not score positive in the antibody test.
2. *Infected individuals who never mount an immune response.* Because the immune response varies from person to person, some infected individuals do not produce antibodies to HIV. There are rare but documented cases of individuals who remain antibody negative but spread HIV infection to their sexual partners.

The HIV antibody test measures whether an individual has circulating antibodies to HIV. However, strictly speaking, the test does *not* indicate if an antibody-positive individual still harbors infectious virus. In principle, some individuals who are exposed to HIV might have raised a successful immune response and completely eliminated the infection. However, by and large, most HIV-antibody positive individuals turn out to be still infected.

Testing for the Level of Circulating HIV

Although the HIV antibody test is routinely used to identify individuals who have been exposed to HIV, other tests for viral infection are used as well. In particular, it is advantageous to know the amount of circulating virus particles in an infected individual because although the levels of virus are generally low in asymptomatic individuals, they frequently rise when full-blown AIDS develops. The first test for virus particles was to measure the level of the major HIV core protein (p24 protein) in the blood. Because this test detects p24 protein by use of an antibody against it, it is often referred to as a test for p24 antigen.

More sensitive tests for HIV infection have been developed. These are based on a technique called *polymerase chain reaction*, or *PCR*, which tests for HIV DNA in infected cells or for HIV RNA in virus particles. The standard PCR test for HIV detects HIV RNA in virus particles. This test is used to detect the very low levels of virus in the blood of asymptomatic individuals. As is discussed in Chapter 5, p. 68, the level of HIV RNA in the blood of infected people (called the *viral load*) has become an important measure for monitoring levels of virus in the blood of infected people as well as the effectiveness of antiviral therapies.

How Does HIV Evade the Immune System?

One of the paradoxes about HIV infection is that most infected individuals contain HIV antibodies, but the disease eventually occurs in most cases, even in the presence of these antibodies. This means that HIV antibodies are unable to prevent onset of AIDS, which may be due to several factors. First, the levels of antibodies raised might be insufficient to block the spread of infectious virus. In addition, antibodies can be produced against different parts of the virus. Only some of these antibodies (*neutralizing antibodies*) can inactivate virus and prevent infection. Finally, several unique features of HIV infection provide the virus with ways to evade the immune system.

High Mutation Rates

The HIV envelope proteins are on the outside of the virus particle, and they are important in attaching the virus to the cell receptor. As such, they are the most important targets for neutralizing antibodies. HIV has an unusually high mutation rate, estimated as one DNA base mutation each time an HIV DNA molecule is made by reverse transcriptase. The consequence of this process is that mutations in the HIV *env* gene occur very frequently, so that the exact amino acid sequence of the envelope proteins changes quite rapidly during successive cycles of infection. Equivalent mutations in the *gag* and *pol* genes generally are not compatible with virus survival. Changes in the makeup of HIV envelope proteins have even been observed over time within the same person. Thus, even though an infected individual may raise neutralizing antibodies to the initial infecting virus, those antibodies may not be able to neutralize subsequent viruses with mutated envelope proteins. Thus, HIV can keep one step ahead of the immune system and continue infection.

Latent States

HIV can establish *latent states* in some cells. In these cells, the viral DNA is maintained but virus proteins are not expressed. As a result, these latently infected cells will not be recognized or attacked by the immune system but will remain as reservoirs for infectious virus. At later times, the virus may be activated from these cells. T_{helper} cells are the major cells that carry latent HIV (see Figure 4-7b). Some of the latently infected T_{helper} cells have extremely long lives, which means that HIV-infected individuals are likely to have these latently infected cells lifelong. In addition, latently infected macrophages contribute to latent HIV infection.

Reactivation of latent HIV from carrier cells may also be important in AIDS progression. Infection of cells carrying latent HIV with certain other viruses, such as herpes simplex or cytomegalovirus, may reactivate the HIV. In addition, other stimuli to the immune system (such as infection with other microorganisms) can result in production of factors that reactivate HIV. These secondary infections may be important cofactors in AIDS progression.

Cell-to-Cell Spread

HIV can carry out infection by cell-to-cell spread. That is, if an HIV-infected cell comes into contact with an uninfected cell, the virus may pass to the uninfected cell directly. Neutralizing antibodies are unable to prevent this process because they can attack virus only when it is outside cells.

These properties of HIV also pose another problem. Vaccines are our front line of defense against most viral infections, as described earlier in this chapter. However, the ability of HIV to evade the immune system means that it will be much more difficult to design an effective anti-HIV vaccine.

Azidothymidine (AZT): The First Effective Drug Treatment in HIV/AIDS

The first successful drug against HIV infection and AIDS was AZT (or zidovudine [brand name Retrovir]). The effectiveness and use of AZT are described in more detail in Chapters 5 and 6. However, let us consider its mode of action here.

AZT is very similar in chemical structure to thymidine, one of the building blocks of DNA. However, when AZT is incorporated in place of thymidine during the DNA assembly process, its structure aborts further DNA assembly. This inactivates any growing DNA molecule that has incorporated AZT. During HIV infection, if AZT is present, HIV reverse transcriptase will readily incorporate it into the viral DNA. This will inactivate the viral DNA. It is important that the enzymes responsible for making the chromosomal DNA of the cell (cellular DNA polymerases) do not efficiently incorporate AZT into DNA. As a result, the cell can continue to grow and make its genetic material, but HIV cannot replicate efficiently if AZT is present. Thus, AZT is a *selective poison* for HIV. It exploits an "Achilles' heel" of the virus—reverse transcriptase. This enzyme plays no role in the uninfected cell, but it is vital to the virus. An agent that affects this enzyme will have no effect on the uninfected cell but will inhibit virus infection. This action is shown in Figure 4-12.

Limitations of AZT

Although AZT is an effective drug in AIDS treatment, it has some limitations:

Toxic side effects. Normal cellular DNA polymerases do not efficiently incorporate AZT into DNA in comparison to HIV reverse transcriptase— the basis for the drug's selectivity. However, cellular DNA polymerases do incorporate some AZT into cell DNA at low levels. During prolonged treatment, this can lead to death of normal cells. *Anemia* is a common side effect in individuals taking AZT; it results from the killing of blood cells by the drug.

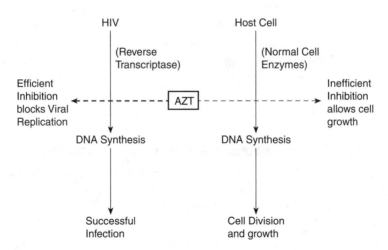

Figure 4-12   The action of AZT.

Inability to halt progression to AIDS. AZT treatment improves the clinical condition of individuals with full-blown AIDS, but it is not a cure (see Chapter 5, p. 87). The inability of AZT to halt progression to AIDS is related to the fact that the virus can mutate in the individual. Indeed, AZT-resistant HIV develops in individuals who have been taking AZT.

Development of AZT-resistant variants. As described earlier, HIV has a high mutation rate. Normally, few mutations in the *pol* gene appear because they decrease the virus's growth rate if they occur. However, under the selective pressure of AZT, AZT-resistant variants of HIV appear that have mutated reverse transcriptase. This mutated enzyme does not incorporate AZT as efficiently, making the virus less sensitive to the drug. AZT-resistant HIV variants develop in AIDS patients who are taking AZT, and these variants result in the inability of AZT to prevent the effects of HIV indefinitely.

Despite its limitations, the effectiveness of AZT in treating AIDS patients has a very important implication. Even in individuals who are already infected, *prevention of continued HIV infection improves the clinical status.* Thus, other drugs that selectively inhibit HIV reverse transcriptase will be useful therapeutic agents. Moreover, the other HIV proteins are all potential "Achilles' heels" for the virus as well. Agents that interfere with the action of any of these proteins may also be useful therapeutic agents. AIDS researchers are devoting a great deal of effort to developing new anti-HIV drugs.

After the initial success of AZT in inhibiting HIV replication and improving the clinical condition of AIDS patients, several other drugs that work by the same general mechanism were developed and approved for use. These include *dideoxycytidine (ddC)*, *dideoxyinosine (ddI)*, *3TC (lamivudine)*, and *d4T (stavudine)*. The active forms of these drugs are preferentially incorporated into viral DNA by HIV reverse transcriptase, and they all prevent further DNA synthesis. As a class, these drugs (including AZT) are referred to as nucleoside analogs, or nucleoside reverse transcriptase inhibitors (NRTIs). Different nucleoside analogs are incorporated into HIV DNA in place of different DNA bases; for instance, ddC is incorporated into HIV DNA in place of the base cytosine (or C).

Protease Inhibitors: Another Class of Drugs Against HIV

In 1996, a new class of anti-HIV drugs became available: *protease inhibitors.* These drugs are targeted on the viral enzyme protease. Earlier in this chapter (p. 49), we learned that protease is important for the conversion of immature virus particles to mature ones (see Figure 4-5). In fact, if this maturation does not take place, the immature HIV particles are not infectious. Thus, drugs that inhibit the protease enzyme will inhibit production of infectious HIV.

A more detailed view of the role of protease in viral maturation is shown in Figure 4-13. When the viral proteins specified by the *gag* gene are initially made in the infected cell, they are in the form of one long chain of amino acids (see Figure 3-7). That is, the amino acids for one gag protein are attached to the next gag protein in one long string, referred to as a *polyprotein*. The same is true of the viral proteins specified by the *pol* gene. In fact, when the virus particles initially form, it is the viral polyproteins that combine with viral RNA to make the immature virus particles that bud from the infected cell. The viral protease liberates the individual *gag* and *pol* gene proteins from the polyproteins by cutting between specific amino acids.

The protease inhibitors work by binding to the HIV protease enzyme and directly inhibiting its function. This differs from the way in which nucleoside analogs such as AZT inhibit HIV replication. Nucleoside analogs do not actually inhibit reverse transcriptase, but they take advantage of the fact that reverse transcriptase will preferentially incorporate them into growing HIV DNA and inactivate it. Several protease inhibitors have been approved for treatment of HIV-infected individuals. They have made a significant improvement in treatment and are discussed in greater detail in Chapter 5.

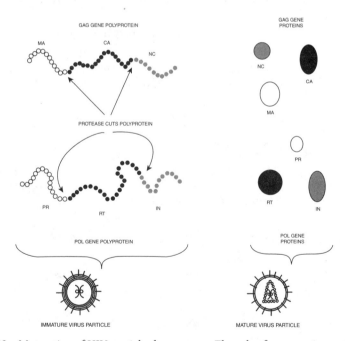

Figure 4-13   Maturation of HIV particles by protease. The role of protease in cutting the *gag* and *pol* gene polyproteins to give the proteins of the mature HIV particle is shown. The abbreviations for the viral proteins are as follows: MA = matrix (p17); CA = caspid (p24); NC = nucleocapsid (p10); PR = protease; RT = reverse transcriptase; IN = integrase. Note that the shape of the core in the mature HIV particle is conical.

Where Did HIV Come From?

Molecular biologists have examined the genetic structure of HIV (actually HIV-1 and HIV-2) in great detail and compared it with the structure of other retroviruses of the lentivirus subclass. From these studies, it is clear that HIV has a common origin with other lentiviruses, and they evolved from a common ancestral retrovirus over millions of years. Retroviruses are ancient viruses, and they have coevolved with their host species. Generally, retroviruses native to a species can replicate in that species without causing disease—a successful adaptation of virus and host. However, when a retrovirus of one species infects another species, it often causes severe disease. In particular, HIV-1 and HIV-2 represent recent infections in humans of lentiviruses native to African primates (*simian immunodeficiency virus*, or *SIV*). HIV-1 came from infection in humans of an SIV from chimpanzees, and HIV-2 came from an SIV of sooty mangabeys. These viruses cause little disease in their natural hosts (chimpanzees and sooty mangabeys), but they cause AIDS in humans. Epidemiological studies tell us that 30 to 40 years ago, HIV-1 spread into high-density populations in Africa and the Western world, leading to the AIDS epidemic. Recent changes in human social behavior, such as the sexual revolution, may have also contributed to the spread of HIV infection.

Numerous apocryphal stories as to the origin of HIV have circulated since the beginning of the AIDS epidemic: HIV was the result of germ warfare research by the CIA; HIV was a laboratory accident involving recombinant DNA; HIV resulted from a plot between Israel and South Africa; HIV resulted from sexual relations between humans and sheep; and HIV resulted from sexual relations between humans and monkeys. *None of these theories are true.*

http://biology.jbpub.com/fan/aids/6e/

Connect to this book's website: http://biology.jbpub.com/fan/aids/6e/. The site features summaries of the main points from each chapter, links to important AIDS-related websites, and short-answer-style review questions for each chapter.

CHAPTER 5

Clinical Manifestations and Treatment of AIDS

In Chapters 3 and 4, we learned about AIDS at the cellular and subcellular levels. In particular, cells of the immune system were discussed, and the effects of HIV infection on those cells were presented. With that background, we can now consider the effects of HIV in terms of a whole person—the actual symptoms that infected individuals experience. A brief overview of AIDS at the organismal level is included in Chapter 4, and in this chapter a detailed description of the clinical manifestations of AIDS is presented. The physical manifestations of the disease are of great importance to AIDS patients and their healthcare providers, but it is important to remember the human side of the disease as well. Untreated AIDS is generally a fatal disease. For persons to learn they are infected with HIV evokes tremendous emotional stress. Counselors are trained to help patients deal with this psychologically difficult situation. In this chapter, we present the biological and clinical aspects of AIDS, as well as current treatments. The psychological consequences of being infected with HIV or learning that one is HIV infected are at least as important and are addressed in Chapter 10.

Exposure, Infection, and Disease

In terms of AIDS and HIV at the organismal level, it is important to consider interaction of the virus with a susceptible individual. Three important concepts in the interaction are *exposure*, *infection*, and *disease*.

Exposure Versus Infection

When an HIV-infected individual encounters an uninfected person, this does not always result in transmission of HIV to the uninfected person. Indeed, even if exposure occurs by one of the three routes known to transmit the virus (blood, birth, and sex), only a fraction of exposed people will be infected. The relative risk factors affecting the efficiency of HIV transmission are discussed in Chapters 6 and 7. As we shall see, different kinds of exposure between infected and uninfected individuals have different probabilities of leading to infection.

As introduced in Chapter 4, most individuals who become HIV infected do not realize they are infected right away, and they remain without symptoms (asymptomatic). It is often not possible to distinguish infected and uninfected people simply on the basis of their physical well-being.

The HIV antibody test is invaluable in identifying individuals infected with HIV (see Chapter 4, p. 58). Generally, an infected person will begin to produce antibodies against HIV (*seroconvert*) 2 to 3 months after infection, although the time for seroconversion is variable and can take up to a year or more to occur. In practical terms, someone who was exposed to HIV is generally considered to be uninfected if he or she is seronegative for HIV antibodies 6 months after the last exposure to HIV and remains seronegative for another 6 months during which no other potential exposures occur.

Infection Versus Disease

Even among individuals who become infected with a virus, not necessarily everyone will develop physical symptoms. With many viruses, most infected individuals never actually develop physical signs of illness. Unfortunately, most people infected with HIV ultimately develop some disease symptoms (see Chapter 6).

The disease symptoms that result from virus infection are caused by destruction or damage of cells and tissues in the infected person. In some cases, the damage may result from direct killing of cells by the infecting virus. In other cases, the physical symptoms may result from indirect effects of the virus. In the case of AIDS, most physical symptoms are the indirect results of damage to the immune system by HIV (see Chapters 3 and 4).

Many virus infections can cause a variety of physical symptoms. Other factors can influence the exact nature of the symptoms in a particular individual, including age, sex, genetic makeup, nutrition, environmental factors, and encounters with other infectious agents. As we shall see, this is particularly true for AIDS, in which the symptoms result from indirect immunological damage.

A schematic diagram of the different clinical stages of HIV infection is shown in Figure 5-1a. In time sequence, the stages can be grouped into three categories: (1) initial infection and the *asymptomatic period*, (2) *initial symptoms*, and (3) *immunological damage* (early signs through full-blown AIDS). A typical time course for events associated with HIV infection is shown in Figure 5-1b.

HIV Infection in Untreated Individuals

We will first consider what happens to someone who becomes infected by HIV and does not receive treatment with antiviral drugs. This situation is rare today in the United States and other more developed countries, but we have extensive information from observation of infected people early in the AIDS epidemic when antiretroviral drugs were not available. Unfortunately, in certain parts of the world, these drugs are still not readily available, so infected people in these areas are still undergoing the same progression of symptoms. Even in the United States, if people become infected but are unaware of it, they will follow the same progression of events described here until they experience some symptom of an AIDS-defining illness.

Primary Infection and the Asymptomatic Period

Many people who become infected with HIV never experience any symptoms at the time of initial infection. On the other hand, some develop relatively mild and transient disease symptoms right after infection (before seroconversion). These are referred to as acute symptoms, and they generally last only a few days and then disappear. Two types of *acute symptoms* can occur.

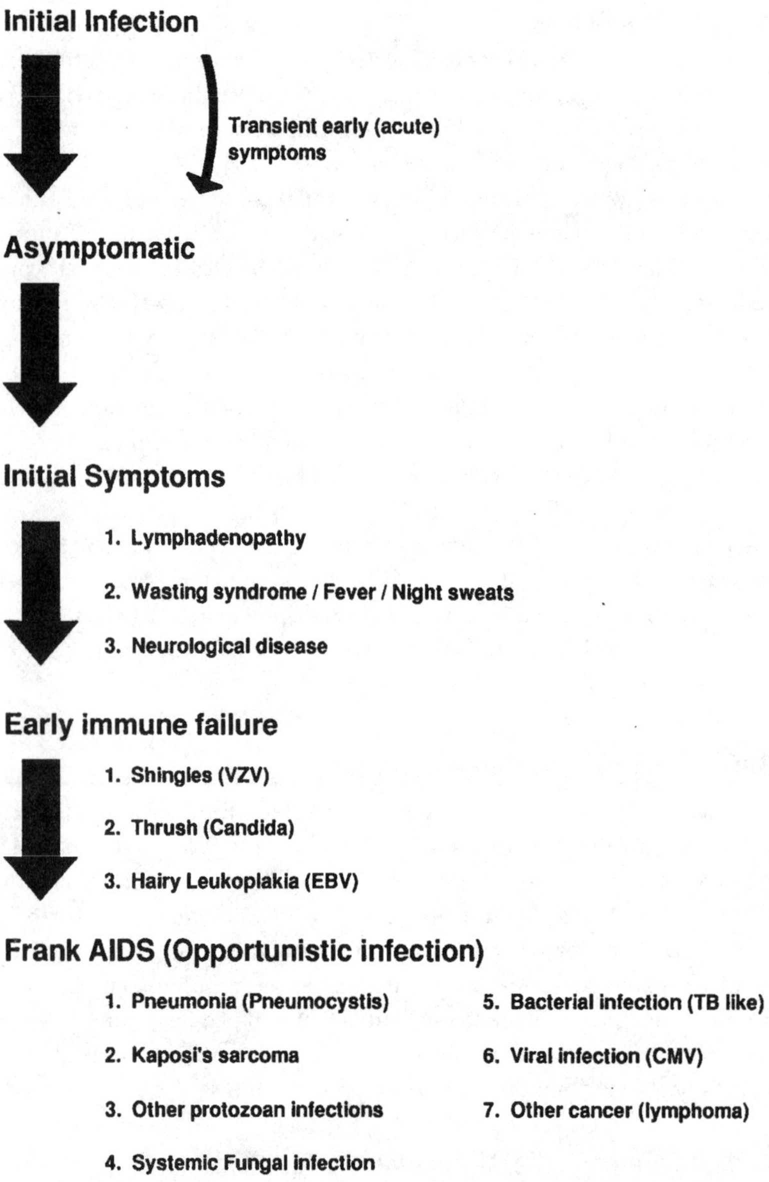

Initial Infection

Transient early (acute) symptoms

Asymptomatic

Initial Symptoms

1. Lymphadenopathy

2. Wasting syndrome / Fever / Night sweats

3. Neurological disease

Early immune failure

1. Shingles (VZV)

2. Thrush (Candida)

3. Hairy Leukoplakia (EBV)

Frank AIDS (Opportunistic infection)

1. Pneumonia (Pneumocystis)

2. Kaposi's sarcoma

3. Other protozoan infections

4. Systemic Fungal infection

5. Bacterial infection (TB like)

6. Viral infection (CMV)

7. Other cancer (lymphoma)

Figure 5-1a   The progression of symptoms in AIDS. (These symptoms may be additive.) EBV = Epstein-Barr virus; TB = tuberculosis; VZV = varicella-zoster virus.

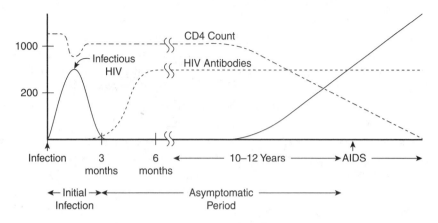

Figure 5-1b   The relative amounts of infectious virus, CD4+ T_{helper} lymphocytes and HIV antibodies circulating in the blood of a typical HIV-infected individual at different times after infection. Note the break in the scale—time to the right of the break is in years, while time before the break is in months.

Mononucleosis-like Illness

The most common early illness seen with HIV infection resembles another viral disease, mononucleosis. Mononucleosis is not exclusive to a particular virus in that several viral infections can cause these symptoms. The most prominent symptoms are swollen lymph glands. In the case of HIV infection, this includes lymph glands throughout the body—called generalized *lymphadenopathy*. In addition, there may also be a sore throat, a fever, and a skin rash. Because these symptoms also result from infection by other viruses, it is not possible to diagnose an HIV infection solely based on the appearance of these symptoms.

Brain Infection (Encephalopathy)

HIV infection of the brain can occur at this early time and lead to brain swelling or inflammation, particularly of the brain lining, or meninges. In medical terms, this condition is called *encephalopathy*. Macrophage cells in the brain appear to be prominent sites for virus replication. The brain inflammation may result from the influx of immune system cells to fight the infection or from the release from infected cells of highly active molecules that can affect other brain cells. The brain inflammation causes symptoms of headache and fever. Often the person will have difficulty concentrating, remembering, or solving problems. Some personality changes may also occur during the acute phase.

The Asymptomatic Period

During the acute phase of infection, quite high levels of circulating infectious HIV are generally detectable in the blood. After the acute phase, the levels of infectious virus in the blood decrease, often to undetectable levels. The infected person usually feels well but becomes *seropositive* for HIV. This is referred to as the period of *asymptomatic infection*. As mentioned earlier, the asymptomatic period may last as long as 10 or more years or sometimes less than 1 year. We do not understand why there is such variability.

During the asymptomatic period, some type of balance is established between HIV infection and the immune system in the infected person. However there is low-level virus replication, which allows the possibility of virus mutations and evolution. Ultimately, for most individuals, changes in the virus or the immune system allow the HIV infection to escape from control and lead to disease.

As described in Chapter 4, p. 61, the level of circulating HIV particles in the blood as measured by the PCR assay is referred to as the viral load. After initial infection, different asymptomatic individuals establish different levels of viral load in the blood that typically are relatively constant throughout the asymptomatic phase. These are referred to as the set points of the viral loads. Within an infected individual, the viral loads will increase above the set point as the individual progresses to AIDS. However asymptomatic HIV-infected people with high viral load set points early after infection progress to clinical AIDS faster than individuals with low set points. As a result, viral load measurements have become important in treating and monitoring asymptomatic HIV-infected individuals. We will return to this later (see p. 89).

Initial Disease Symptoms

The initial diseases that follow the asymptomatic period fall into three major classes. An infected person may have symptoms of more than one of these classes.

Wasting Syndrome

The two symptoms seen with this syndrome are a sudden and otherwise unexplained *loss in body weight* (more than 10% of total body weight) and fevers, usually at night, that cause *night sweats*. The weight loss is usually progressive, leading to wasting away of the infected person, and may be accompanied by diarrhea. This wasting syndrome is very reminiscent of the progressive loss of body weight by cancer patients. The fevers can involve dangerously high temperatures (106–107°F), which can result in brain damage. Normally, the body controls high internal temperatures by sweating. The night sweats result from the bodies of infected individuals lowering their temperatures by sweating.

Lymphadenopathy Syndrome

As described in the section on mononucleosis-like symptoms during initial infection section, *lymphadenopathy* means swelling of the lymph glands. Lymphadenopathy sometimes is also an acute symptom of HIV infection, but in lymphadenopathy syndrome (LAS), the lymph gland enlargement is persistent. This condition is also called *persistent generalized lymphadenopathy*. In LAS, the lymph glands in the head and neck, the armpits, and the groin are usually swollen, although they generally are not painful. Some infected people experience both LAS and wasting syndrome. In the past, lymphadenopathy was one of a group of symptoms that was associated with AIDS-related complex (ARC), a condition considered less severe than AIDS. However, the term *ARC* is not used today, and lymphadenopathy actually can occur at various stages of the disease.

Neurologic Disease

The HIV infection can spread to the brain and either damage the brain directly or lead to damage by other infectious agents. In addition, other parts of the nervous system can be damaged and cause different neurologic symptoms. These symptoms can occur early or late in progression toward AIDS. About one-third of all AIDS patients have some of the following neurologic symptoms:

Dementia. When the brain itself is damaged, mental functions are impaired. With HIV infection, this is usually a progressive situation. Initially, dementia may appear as simple forgetfulness about where things are. As the disease progresses, the loss of mental function can become more serious: The infected person may have difficulty reasoning and performing other mental tasks. Depression, social withdrawal, and personality changes are also common. Eventually, as the disease progresses, infected people may become unable to care for themselves. For some AIDS patients, this progression leads to the patient entering a coma followed by death, if other infections or cancers do not kill the patient first. Death usually occurs several months after the onset of dementia.

Spinal cord damage (myelopathy). Because the spinal cord transmits nerve impulses to the muscles of the body, damage to the spinal cord can result in weakness or paralysis of voluntary muscles. As a result, HIV infection can lead to spinal cord swelling (*myelopathy*) and paralysis or weakness of the limbs.

Peripheral nerve damage (neuropathy). Some people infected with HIV experience swelling (neuropathy) of the peripheral nerves. These nerves are involved in sensing pain. When they are damaged, they can cause burning or stinging sensations, usually in the hands or feet. In addition, numbness, especially in the feet, is frequent.

These initial symptoms of HIV infection are not mutually exclusive. Individual patients may experience a mixture of any of them.

Damage to the Immune System and Full-Blown AIDS

As described in Chapter 4, the major problem in HIV infection is damage to the immune system. Two major consequences result from immunological damage: the occurrence of *opportunistic infections*, caused by infectious agents that are normally held in check by healthy immune systems, and the development of *cancers*, which also result from failure of the immune system (see Chapter 3, p. 34). In HIV-infected individuals, both opportunistic infections and cancers may develop (sometimes at the same time), but opportunistic infections are generally the more common cause of death.

As indicated in Chapter 4, the breakdown of the immune system in HIV-infected individuals is a continuous and gradual process. It generally begins with relatively minor opportunistic infections and usually progresses to severe and life-threatening disease—that is, full-blown AIDS. Opportunistic infections associated with early immune failure can become established in individuals whose immune systems are still partially intact, while those infections associated with full-blown AIDS will only become established when the immune system is severely damaged. After the isolation of HIV, an early medical definition for diagnosing AIDS was evidence of HIV infection (seropositivity) in conjunction with two or more serious opportunistic infections or cancers (termed AIDS-defining illnesses). This definition was initially used by epidemiologists and clinicians in staging and classifying the disease. However, we now know that AIDS represents the final and most severe condition resulting from HIV infection, and it is not really distinct from other manifestations of HIV disease. In 1993, a new criterion was added to the AIDS definition: evidence of HIV infection (seropositivity) and T_{helper} counts below 200 per cubic millimeter in the blood. In adults, full-blown AIDS almost never occurs before 2 years of infection.

Early Immune Failure

Several opportunistic infections occur in HIV-infected individuals with moderate immune deficits. Following are some of the more common opportunistic infections that occur during early immune failure.

Candida. *Candida* is a genus of fungus, similar to baker's yeast, that can be found on the skin and mucosal surfaces (mouth, vagina) of most people. Normally, *Candida* growth is held in check by an ecological balance with other microorganisms and by the immune system. In HIV-infected individuals with moderate immununodeficiency, *Candida* often infects the mouth, causing a condition known as oral *candidiasis*, or thrush. With thrush, the *Candida* will form white plaques in the mouth that feel furry to the patient (Figure 5-2). Antifungal drugs such as Mycostatin are used to control these infections, although they are difficult to completely eliminate.

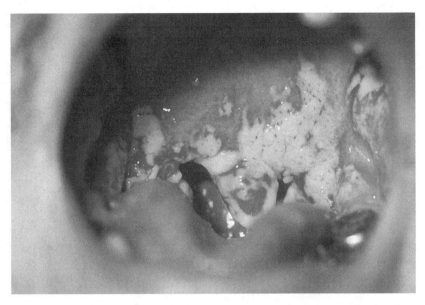

Figure 5-2   Oral candidiasis. The photograph shows the inside of the mouth. The white spots are areas of *Candida* (yeast) infection. (Courtesy of CDC.)

HIV-infected people who develop candidiasis have a high probability of progressing to full-blown AIDS. With more severe immunodeficiency, the infection can spread down the esophagus and cause a very painful burning sensation when the patient eats. This condition is known as *esophagitis*; patients with esophagitis are generally considered to have full-blown AIDS. Approximately 50% of AIDS patients will experience, at some time, a *Candida* infection. In HIV-infected women, vaginal *Candida* infections are a very important symptom.

Shingles (varicella). Shingles, or *varicella*, is a painful rash condition that often occurs on the torso (Figure 5-3). It is caused by the reactivation of a latent virus called *varicella-zoster*. This is the virus that causes chicken pox during childhood; it is a member of the herpesvirus family. After the initial childhood infection, the virus can remain dormant in the nerve trunks for many years and become reactivated when the immune system is compromised or stressed. With AIDS patients, the severity of shingles appears to be greater than that seen in non-AIDS patients, presumably because of their failing immune systems. The antiviral drug acyclovir is used to help control shingles.

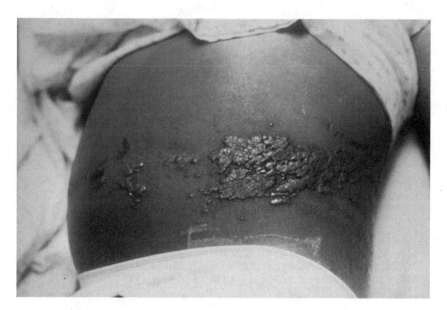

Figure 5-3 Shingles. Reactivation of the latent varicella-zoster (chicken pox virus) infection. (Courtesy of CDC.)

Hairy leukoplakia. This is an abnormal condition of the mouth in which white plaques appear on the surface of the tongue. These plaques are not caused by the overgrowth of a fungus or bacteria, however. They are the abnormal growth of the papillae cells of the tongue; these plaques cannot be scraped off. These overgrown cells resemble cancer cells and result from infection with another virus called *Epstein-Barr virus.* Epstein-Barr virus is also a member of the herpesvirus family and causes infectious mononucleosis in young adults. Hairy leukoplakia is a condition unique to AIDS patients.

Tuberculosis. Tuberculosis is caused by bacterium of the *Mycobacterium* genus. Mycobacteria have characteristic cell walls with unusual staining properties in the laboratory. Tuberculosis can occur in HIV-infected individuals with early immune failure or when they have more severe immunodeficiency. Tuberculosis has become a common infection in HIV-infected individuals and AIDS patients. Tuberculosis was largely eradicated in the United States by the mid-1970s through public health measures and antibiotic treatments. However, recently there has been a resurgence of tuberculosis due to a combination of several factors: (1) increased immigration from areas where tuberculosis infection is still common (such as developing countries in Asia and Latin America); (2) a decline in funding for public health measures, which has allowed tuberculosis to spread; and (3) HIV/AIDS patients, who are highly susceptible to tuberculosis infection and who in turn can transmit the infection. Tuberculosis is typically an infection of the lungs, although the bacteria can infect internal organs such as the bone marrow as well. The antibiotics used in treatment of tuberculosis include isoniazid and rifampin.

A very disturbing trend is the increasing appearance of tuberculosis strains that are resistant to the standard antibiotics. This resistance probably results from incomplete therapy of tuberculosis patients: To eradicate the bacteria, a lengthy course of antibiotics is required, with repeated follow-up and testing until the bacteria are completely eliminated. However, if follow-up is incomplete, even if most of the tuberculosis bacteria are eliminated, the surviving bacteria can reestablish and spread to other individuals after antibiotic therapy is stopped. In addition, these bacteria are frequently resistant to the antibiotic that was administered. Tuberculosis is now a major threat to healthcare workers, particularly in hospital settings where there are large numbers of indigent individuals, including HIV-infected individuals and AIDS patients.

AIDS

As discussed elsewhere in this book (Chapter 4, p. 57, and Chapter 6), without antiviral treatments most HIV-infected individuals will develop some of the symptoms associated with AIDS within 10 to 12 years after initial infection. The rate at which infected individuals develop symptoms may vary somewhat among different risk groups. For instance, hemophiliacs who were infected by transfusions or blood products may develop AIDS at a slower rate than do homosexual men who had high numbers of sexual partners. This may be influenced by the number and nature of other microorganisms (potential opportunistic infections) that these people encounter.

The following infections and cancers seen in AIDS patients are indications that the immune system has undergone a catastrophic failure and can no longer prevent life-threatening infections or cancers. They typically occur in HIV-infected individuals with CD4 cell counts of less than 200/cubic millimeter of blood.

■ Fungal Infections

Pneumocystis pneumonia (PCP). This illness, which results from inflammation of the lungs, was originally the most common of the serious secondary infections seen with AIDS, and it has been a leading cause of death in AIDS patients. Before the development of antiretroviral drugs, 70–80% of AIDS patients developed PCP; 20–40% of these individuals would die. Inflamed areas of the lungs appear as white spots in x-rays (Figure 5-4a). The inflammation is caused by infection with a fungus called *Pneumocystis jirovecii* (Figure 5-4b). (Until recently, this microorganism was often classified as a protozoan, but it is now considered a fungus, based on molecular biological studies. The *Pneumocystis* that infects humans has recently been re-named *Pneumocystis jirovecii*; it was previously called *Pneumocystis carinii*, which now only refers to the species that infects rodents.) This microorganism gets its name from the fact that it can grow into cysts in the lungs of rats (pneumocystis). *P jirovecii* is relatively common, and small numbers of the fungus can be found in the lungs of healthy people, and the related microorganism *P carinii* is found in many animals. In AIDS patients, the infection is often insidious, and the patient may be unaware of the seriousness of his or her illness. A dry cough is common, and a progressive shortness of breath indicates poor lung function. The shortness of breath is due to the inability of the inflamed lungs to take up adequate amounts of oxygen, which can lead to tissue damage throughout the body. *P jirovecii* particles are detected by staining the fluid washed out of the lungs with a special dye. PCP can be treated with various antibiotics called *sulfa drugs*. The antiparasitic agents trimethoprim and sulfamethoxazole (TMP-SMX) are usually given together (as the drug Bactrim) to control the infection. Another drug, called *pentamidine*, is used when TMP-SMX becomes toxic to the patient.

Systemic mycosis. Three types of common soil fungi can cause generalized infections in AIDS patients. These fungi can exist in either moldlike or yeastlike forms and are called dimorphic. The three types are *histoplasma* (causing histoplasmosis), *coccidian* (causing coccidiomycosis), and *cryptococcus* (causing cryptococcal meningitis). These fungi can cause lung infections in healthy people, but generalized or systemic infections are very rare. In AIDS patients, these fungi can cause devastating systemic infections that are massive and widespread. The brain, skin, bone, liver, and lymphatic tissue may all be highly infected, typically leading to death of the patient. Antifungal drugs such as miconazole are used to control these infections.

■ Protozoal Infections

Cryptosporidium *gastroenteritis.* This disease is caused by a protozoan called *Cryptosporidium*. This protozoan infects the linings of the intestinal tract and causes diarrhea (gastroenteritis). In healthy people, diarrhea from a *Cryptosporidium* infection is normally mild, lasting only a few days. However, in AIDS patients the diarrhea is prolonged and severe. The AIDS patient may have from 20 to 50 watery stools per day, accompanied by abdominal cramps and profound weight loss. As a result, there is a serious loss of fluid and electrolytes (salts in the blood). Patients are treated with

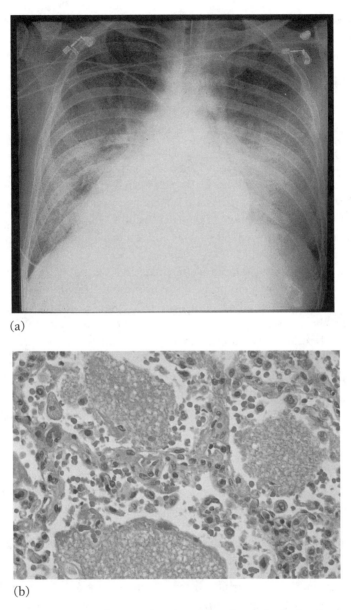

(a)

(b)

Figure 5-4 *Pneumocystis* pneumonia. (a) A chest x-ray of an individual with PCP is shown. The ribs are apparent in the top of the x-ray, but they are not clear in the bottom. This is because the lungs inside the rib cage are filled with fluid, and they make that area of the x-ray appear lighter. In a normal individual, the ribs would be evident against a clear background at the bottom of the picture as well. (b) Lung tissue from an individual with PCP. The dark round particles are *Pneumocystis* microorganisms within the lung tissue. (Part a courtesy of Judith Feinberg, MD; part b courtesy of Dr. Edwin P. Ewing, Jr./CDC.)

intravenous fluids and electrolytes, and their diarrhea can be controlled somewhat with drugs that slow intestinal action. Currently no standard antibiotic is recognized for use against *Cryptosporidium*. Only about 5% of AIDS patients develop this disease. *Cryptosporidium* also infects cattle and other animals, especially their young; these animals may be the source of human infection. Contaminated drinking water (e.g., through travel in developing areas with poor sanitation) is one source of *Cyptosporidium* infection.

Toxoplasmosis. This disease is caused by a species of protozoan called *Toxoplasma gondii*, which normally causes an asymptomatic infection in healthy adults. This protozoan also infects a very wide variety of animals; domestic cats are one source of human infection. Unlike *Cryptosporidium*, *Toxoplasma* is an intracellular parasite and can invade numerous organs of infected individuals.

In AIDS patients, the brain is often infected (*Toxoplasma* encephalitis), which may result in symptoms similar to those seen with brain tumors: convulsions, disorientation, and dementia (Figure 5-5). A computed tomography (CT) scan is used to diagnose toxoplasmosis. *Toxoplasma* infection in humans is quite common, and it normally does not result in disease. The *Toxoplasma* encephalitis in AIDS patients represents reactivation of already existing latent *Toxoplasma* organisms.

Various antibiotics, such as pyrimethamine and sulfadiazine, are effective treatments, but they must be administered indefinitely to prevent a relapse. Unfortunately, some patients develop toxic reactions to these drugs.

■ Bacterial Infections

Interestingly, infections by commonly occurring bacteria (such as those in the lower intestines) do not generally occur in adult AIDS patients, perhaps because the components of the immune system responsible for controlling the common bacteria are less affected by HIV infection. However, children born infected with HIV often develop lung infections with common bacteria. In addition, adult AIDS patients may experience infections with tuberculosis-like bacteria.

Mycobacterium. As described in the Early Immune Failure section, tuberculosis is caused by a mycobacterium. AIDS patients with severe immunodeficiency (typically with less than 25 CD4 cells/cubic millimeter of blood) also develop infections with an atypical form of tuberculosis bacterium called *Mycobacterium avium-intracellulare*. This bacterium does not normally cause disease in healthy people, but in AIDS patients it may cause a tuberculosis-like disease in the lungs. The infection can also involve

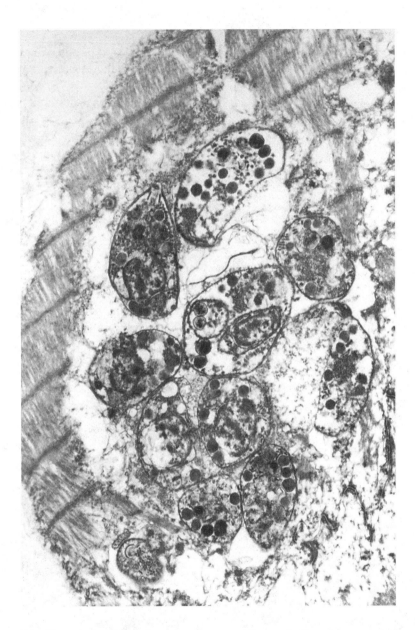

Figure 5-5   Toxoplasmosis. An electron microscope picture of a *Toxoplasma* cyst is shown. A cyst is a walled-off area of microorganisms within a tissue. Each of the round areas in the photograph is a cross-section of a *Toxoplasma* microorganism. (Courtesy of Dr. Edwin P. Ewing, Jr./CDC.)

numerous other tissues, such as the bone marrow, and bacteria may be present in the blood at very high levels (Figure 5-6). Isoniazid and rifampin are usually among the drugs used. This infection is more common in AIDS

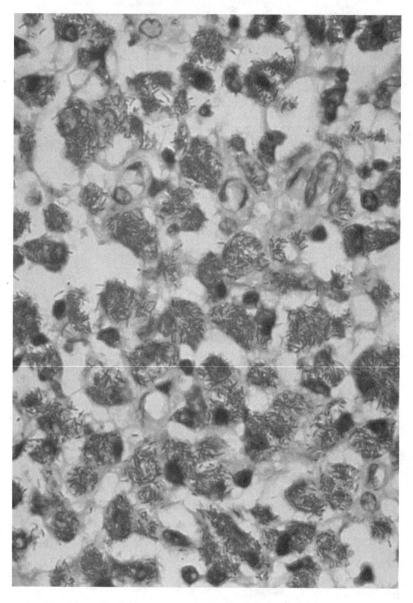

Figure 5-6   *Mycobacterium avium-intracellulare* infection of lymph node in a patient with AIDS. The small, dark, rod-shaped particles are the mycobacterium. (Courtesy of Dr. Edwin P. Ewing, Jr./CDC.)

patients who were injection drug users. As better antiviral treatments have been developed that more effectively control CD4 cell loss, the frequency of atypical mycobacterium infections in AIDS patients has declined.

Viral Infections

The viral infections that cause symptoms in AIDS patients result from viruses that can establish latent infections, as was described for shingles and varicella-zoster virus in early immune failure. It is interesting that other viruses (e.g., the virus that causes the common cold) do not cause severe disease in AIDS patients.

Cytomegalovirus. This virus is a member of the herpesvirus family, as are the varicella-zoster and Epstein-Barr viruses described earlier. Cytomegalovirus (CMV) is a common virus, and many people are infected early in childhood. Children tend to get an asymptomatic infection, whereas infected young adults may develop a mononucleosis-like illness. Infection of a fetus (a congenital infection) is very serious and can lead to permanent brain damage or death of the fetus. In AIDS patients, CMV infection can recur and tends to infect the retinas of eyes, leading to blindness. The virus also infects the adrenal gland, leading to hormonal imbalance. Pneumonia, fever, rash, and gastroenteritis due to CMV infection are also seen in AIDS patients. CMV pneumonia in patients who have PCP at the same time is usually fatal. The antiviral drug ganciclovir (related to acyclovir) is used to control CMV infections.

Herpes simplex virus (HSV). HSV is another member of the herpesvirus family. This is a common viral infection, and like other herpesviruses, HSV can establish latent infections. There are actually two HSVs: HSV-I infects cells of the oral epithelium and causes cold sores or fever blisters around the mouth; HSV-II infects epithelial cells of the reproductive tract and causes sores or blisters of the male and female reproductive tracts (genital herpes infections). In AIDS, patients' latent HSV infections can recur (activate) with increased frequency, leading to persistent and painful sores in the reproductive tract. HSV infections and reactivations are treated with the antiviral compound acyclovir.

Hepatitis viruses. Hepatitis viruses cause inflammation of the liver. There are actually several different hepatitis viruses, all of which infect the liver and cause inflammation and acute liver damage. Two of these viruses, *hepatitis B virus (HBV)* and *hepatitis C virus (HCV)*, can establish persistent infections in some individuals in which the virus is not eliminated. The persistent infection can eventually lead to chronic liver destruction (chronic active hepatitis) and scarring (cirrhosis)—serious conditions. Individuals who develop chronic active hepatitis also have an increased likelihood of developing liver cancer (hepatocellular carcinoma). Individuals who are infected with both HIV and either HBV or HCV can experience more severe liver disease (cirrhosis or liver cancer from HBV or HCV) as the immune system declines (from HIV). HCV infection is common in injection drug users, and HBV infection can result from either injection drug use, sexual contact, or perinatal transmission (from mother to child). Management of the hepatitis virus infections in HIV-infected individuals is challenging. Some antiretroviral drugs, such as the NRTI lamivudine, are effective against HBV because this virus shares some features with retroviruses—including a viral reverse transcriptase. HCV is quite difficult to treat. The current therapies, which include combined treatments with the drugs interferon and ribavirin, are not effective against all HCV strains.

JC virus and nervous system damage. JC is a small DNA-containing virus of humans. The virus has a preference for infecting cells of the nervous system, and in most individuals the infection is controlled by the immune system. Nevertheless low levels of latent or persistent infection can be detected in many people. In AIDS patients, in the absence of an effective immune system, JC virus can vigorously reproduce in brain cells, leading to destruction of brain tissue, a condition known as *progressive multifocal leukoencephalopathy (PML)*. PML is rare, but when it occurs it is usually fatal. There are no treatments for PML, with the exception of restoring the immune system with antiretroviral drugs to the point that the immune system can control the JC virus infection. Unfortunately, damage to the brain resulting from JC virus/PML is irreversible; once brain tissue is destroyed by JC infection, it is not replaced or regenerated.

■ Cancers

As mentioned in Chapter 3, the immune system also defends us from cancers (immunological surveillance) since tumors typically express "foreign" proteins. Thus in AIDS, when the immune system is crippled, cancers can more readily develop. In fact, the cancers characteristic of AIDS are all caused by viruses. A minority of human cancers are caused by viruses. The fact that AIDS patients develop virus-related cancers may actually reflect increased expression of the underlying cancer virus in immunodeficient individuals. AIDS patients do not show higher frequencies of other cancers where there is no viral involvement—e.g., colon cancer, breast cancer, and lung cancer. As mentioned earlier, liver cancer secondary to HBV or HCV infection is observed in AIDS patients who are also infected with those viruses.

Kaposi's sarcoma. Kaposi's sarcoma is a tumor of the blood vessels (Figure 5-7). In non-AIDS patients, Kaposi's sarcoma is typically only seen in older men of Mediterranean or Jewish ancestry. In homosexual men with AIDS

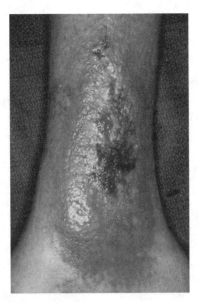

Figure 5-7   Kaposi's sarcoma. Dark (purplish) areas of Kaposi's sarcoma are shown on the skin of an AIDS patient. (Courtesy of Dr. Steve Kraus/CDC.)

who are not treated with antiviral drugs, up to 69% may develop Kaposi's sarcoma. Initially, only a few tumors appear as pink, purple, or brown skin lesions, usually located on the arms or legs. These tumors spread and become widely distributed, eventually involving most of the linings of the body. If they spread to the lungs, they are difficult to control. Chemotherapy can eradicate these tumors with a high success rate. Triaziquone, actinomycin D, bleomycin, and ICRF-159 are often used in chemotherapy. AIDS patients with Kaposi's sarcoma often have a high level of opportunistic infections. Kaposi's sarcoma results from underlying infection with another recently discovered member of the herpesvirus family (human herpesvirus-8 [HHV8], or Kaposi's sarcoma herpesvirus [KSHV]).

Lymphomas. The lymphomas that occur in AIDS patients are cancers derived from the B-cells of the immune system. These are cells that make antibodies, as discussed in Chapter 3. Reactivation or co-infection in the B-lymphocytes with Epstein-Barr virus is involved in development of the lymphomas. As mentioned earlier in this chapter, the Epstein-Barr virus causes mononucleosis in young adults, but it can also transform normal B-cells into cancer cells. In AIDS patients, an unusual lymphoma that spreads to the brain also occurs.

Lymphomas associated with HHV8/KSHV are also occasionally observed in AIDS patients: pleural effusion lymphomas and multicentric Castleman's disease.

Cervical cancer. In female AIDS patients, cancer of the cervix (a part of the female genital tract) is also observed with high frequency. Cervical cancer is a fairly common cancer in women, although it typically affects women of middle age or older. Infection with certain strains of human papillomavirus (HPV) that cause warts in the genital tract is an underlying cause of cervical cancer. Like HIV, genital infection by HPV occurs by sexual contact. Thus, in AIDS patients, this HPV-induced cancer develops more rapidly in the absence of a normal immune system. In male AIDS patients, HPV infection can also give rise to cancers of the male reproductive tract, such as cancer of the penis.

This list of opportunistic infections and cancers seen in AIDS patients covers only the most commonly observed diseases. Numerous other infections are also seen at lower frequencies. In addition, an individual patient may experience a combination of these illnesses. It is interesting that there are characteristic cancers and opportunistic infection in AIDS patients. These diseases are only a fraction of potential diseases that could affect an immunocompromised person. This may be due to the fact that HIV more seriously damages certain parts of the immune system. In addition, other factors, such as previously established chronic or latent infections, may be important. Once the immune system fails, the preexisting infectious agents can then proliferate and cause disease symptoms.

Different groups of HIV-infected individuals may differ in their symptoms. This may result from differing exposures to opportunistic agents, different ages, different sexes, or different geographical locations. For instance, in contrast to homosexual men, women with AIDS have a higher incidence of severe vaginal or oral yeast infections and cervical cancer; on the other hand, they rarely develop Kaposi's sarcoma. Children with AIDS suffer from more bacterial infections than do adult AIDS patients.

High Turnover of T_{helper} Cells in AIDS Patients

As described above, the immunological deficiency in AIDS patients results from the very low numbers of T_{helper} cells present. The low number of T_{helper} cells results from continual high-level destruction by viral infection. Even in an AIDS patient with very low numbers of T_{helper} cells, the immune system produces many T_{helper} cells every day. However, the production of new T_{helper} cells cannot keep up with the destruction of T_{helper} cells brought on by HIV. It is estimated that an AIDS patient loses about 2 billion T_{helper} cells every day and that his or her immune system is making almost the same number every day in an attempt to replenish them. The fact that T_{helper} cells are being produced in AIDS patients (even if they are being destroyed by HIV) has another consequence. If virus replication can be blocked, then it is possible for the T_{helper} cells to build back up.

Antiviral Drugs for HIV

To restore the health of an AIDS patient, it is necessary to suppress replication of HIV and to rebuild the damaged immune system (see Chapter 12, p. 205). So far, antiviral drugs have played the most important part in the clinical treatment of this disease.

Reverse Transcriptase Inhibitors

Reverse transcriptase inhibitors are drugs that target the viral enzyme reverse transcriptase. As described in Chapter 4, reverse transcriptase is necessary for HIV and other retroviruses to replicate, and it is not required for normal cells to function. Therefore, drugs that inhibit reverse transcriptase will specifically block the virus from replicating while they will have little effect on the host cell.

AZT and Other Nucleoside Analogs

AZT was the first drug to show benefit in treatment of AIDS patients, and it is still one of the primary drugs in use (see Chapters 4 and 6). AIDS patients who were given AZT showed increased survival. Without AZT or other antiviral drugs, the average life expectancy of an AIDS patient with an AIDS-defining opportunistic infection was about 6 months. With AZT treatment alone, that life expectancy rose to 1.5 years. Treatment with AZT actually resulted in some recovery of immune function. The number of T_{helper} lymphocytes in AZT-treated AIDS patients increased; they experienced fewer opportunistic infections, and they actually gained weight. The patients also felt better. Although AZT is rarely used by itself anymore, we will first discuss its role as a "monotherapy," because this was the original way it was used (its role in combination therapies will be discussed in the Antiretroviral Therapy section later in this chapter).

The cost of AZT is high—about $3,500 per year for one person. The original treatment regimen demanded that a patient take an AZT pill once every 4 hours, around the clock. As mentioned in Chapter 4, AZT treatment has some side effects, such as nausea, headache, and loss of sleep. The major complication is that about half of treated patients become anemic—that is, they have low white blood cell counts. Anemia itself can lead to an increase in bacterial infections. If anemia occurs, AZT treatment must be discontinued, at least temporarily.

As mentioned in Chapter 4, AZT is a *nucleoside analog*, or *nucleoside reverse transcriptase inhibitor (NRTI)*. Seven other nucleoside analog drugs have also been approved for treatment of HIV-infected individuals and AIDS patients: *dideoxyinosine* (*ddI*), *dideoxycytidine* (*ddC*), *3TC* (*lamivudine*), *d4T* (*stavudine*), *abacavir, tenofovir,* and *FTC* (*emtricitabine*) (Table 5-1). All these drugs are analogs of DNA building blocks that terminate further DNA synthesis if incorporated into DNA by HIV reverse transcriptase. Side effects are also problems with these drugs. For instance, some individuals taking ddI may develop pancreatitis (a serious inflammation of the pancreas), and some individuals on ddC experience peripheral neuropathies (tingling in the extremities).

Table 5-1	Currently Approved HIV Antiviral Drugs	
Drug Class	Chemical/Generic Name	Trade Name
Nucleoside inhibitors (NRTIs)	AZT/azidothymidine/zidovudine	Retrovir
	ddI/dideoxyinosine/didanosine	Videx
	ddC/dideoxycytidine/zalcitabine	HIVID
	3TC/lamivudine	Epivir
	d4T/stavudine	Zerit
	Abacavir	Ziagen
	FTC/coviracil/emtricitabine	Emtriva
Non-nucleoside reverse transcriptase inhibitors (NNRTIs)	Nevirapine	Viramune
	Delavirdine	Rescriptor
	Efavirenz	Sustiva
Protease inhibitors	Saquinavir	Fortovase
	Indinavir	Crixivan
	Ritonavir	Norvir
	Nelfinavir	Viracept
	Amprenavir	Agenerase
	Atazanavir	Reyataz
	Darunavir	Prezista
	Fosamprenavir	Lexiva
	Lopinavir ·	Kaletra
	Tipranavir	Aptivus
Integrase inhibitors	Raltegravir	Isentress
Fusion inhibitors	Enfuvirtide/T-20	Fuzeon

*In addition, some combination pills are available that combine two or three of these drugs.

Development of Drug-Resistant HIV

As discussed in Chapter 4, p. 64, development of drug resistance in HIV-infected individuals taking antiviral drugs is the major limitation of their effectiveness. HIV has an unusually high mutation rate. During the many cycles of viral replication in an infected individual, a mutated virus with an altered reverse transcriptase will be produced that does not efficiently incorporate AZT into HIV DNA. As a result, in HIV-infected individuals who have been taking AZT, there is a selection process for AZT-resistant virus. Once AZT-resistant HIV becomes the predominant virus in the infected individual, AZT treatment does not provide any benefit.

The viral load assays described in Chapter 4 allow us to see how serious the problem of drug resistance is for HIV antivirals. When an HIV-infected individual first begins to take AZT, there is usually a 10- to 100-fold drop in the amount of HIV RNA circulating in the blood. However, within a year, the amount of viral RNA in the blood returns to nearly the same level as before drug treatment began. Moreover, the HIV is AZT resistant.

Drug resistance has been a problem with every HIV antiviral drug developed so far. Treatment with all of the nucleoside analog compounds leads to development of drug-resistant virus (with drug-resistant reverse transcriptase). However, the changes in amino acid building blocks that lead to an AZT-resistant reverse transcriptase may be different from changes in reverse transcriptase resistant to another drug (e.g., 3TC). As a result, HIV that is resistant to AZT is often sensitive to inhibition by 3TC. Thus, long-term treatment of HIV-infected individuals involves rotating them onto new antiviral drugs as their virus becomes resistant to those they are currently taking. Continued development of more HIV antiviral drugs will be important.

AZT Treatment of Pregnant Women

AZT treatment alone has had an important effect on preventing transmission of HIV infection from pregnant women to their children. HIV-infected pregnant women have about a 23% chance of transmitting infection to their offspring. Treating women with AZT during pregnancy lowers the risk for transmitting infection to the newborn by 70%. This is one of the few currently recommended uses of AZT monotherapy.

Non-nucleoside Inhibitors of Reverse Transcriptase

Other drugs that target the HIV reverse transcriptase are also available. In particular, drugs that bind to the reverse transcriptase enzyme and prevent it from functioning have been developed. These are called *non-nucleoside reverse transcriptase inhibitors* (*NNRTIs*) (see Table 5-1). One such drug, nevirapine, was approved for treatment in 1996. Nevirapine has some advantages: It has relatively few side effects, and it leads to even better initial decreases in viral RNA load than do the nucleoside analogs—1,000- to

10,000-fold decreases. However, nevirapine-resistant HIV arises quickly in infected individuals who are taking the drug. In 1997, a second non-nucleoside reverse transcriptase inhibitor (delavirdine) was approved for use, and a third (efavirenz) was approved in 1998.

Recently, a single dose of nevirapine has been shown to be effective in substantially reducing HIV transmission from pregnant women to their children. This is much cheaper and simpler than AZT monotherapy (see preceding section) and is particularly promising for the developing world, where continual treatment of pregnant women with AZT throughout their pregnancy may not be affordable.

Protease Inhibitors

Protease inhibitors are another important class of HIV antiviral drugs. The role of protease in maturation of virus particles is discussed in Chapter 4, p. 64. Protease inhibitors bind to the HIV protease and inhibit its activity. Ten HIV protease inhibitors have been approved for use; the first ones available were saquinavir, indinavir, ritonavir, nelfinavir, and amprenavir (Table 5-1). The protease inhibitors have some advantages when given to patients: very strong initial decreases in viral load (1,000- to 10,000-fold decreases in circulating viral RNA) and relatively few side effects. However, in patients receiving only a protease inhibitor, resistant HIV develops rapidly. Moreover, the drug-resistant virus often shows cross-resistance to other protease inhibitors. For instance, HIV that is resistant to indinavir is generally resistant to ritonavir as well. Thus, once an HIV-infected person taking a protease inhibitor develops drug-resistant virus, it is likely that other protease inhibitors will be of limited benefit. The protease inhibitors have been most effective when used in combination therapies.

Integrase Inhibitors

The most recently developed antiviral drugs target the viral integrase—the enzyme responsible for integration of viral DNA into the host cellular DNA. One integrase inhibitor, raltegravir, was approved for use in 2007, and a few others are undergoing testing. Raltegravir is effective at controlling viral loads when used in combination with other anti-HIV drugs.

Fusion Inhibitors

In 2003, a new antiviral drug that targets an early step in the infection process was approved, enfuvirtide (Fuzeon). After an HIV particle binds to a CD4-containing cell that also expresses a co-receptor (CXCR4 or CCR5), the next step in infection involves fusion of the viral envelope with the plasma membrane of the cell. This fusion allows passage of the HIV core into the cytoplasm of the cell. Enfuvirtide blocks the fusion process, so HIV cores cannot enter the cell, and the infection process is blocked at that step.

Clinical Management of HIV-Infected Individuals

Once individuals are identified as being HIV infected, they should be managed by a doctor. There are two important aspects to management: (1) monitoring the viral infection and disease, and (2) treatment with antiviral drugs to control the HIV infection.

Monitoring Infection and Disease

Because HIV is a chronic infection and it takes years before clinical AIDS develops, it is important to monitor the level of viral presence/infection and the state of the immune system in infected individuals. The primary measure of the immune system status is the T_{helper} cell count in the circulation. This is determined by staining blood cells for the CD4 surface protein (the predominant CD4-positive cells in the blood are T_{helper} lymphocytes) and counting their numbers. As described earlier, normal healthy individuals have greater than 1,000 CD4 cells per cubic millimeter of blood, while individuals with CD4 counts of less than 200 are defined as having AIDS. When CD4 counts are below 400, there is some level of immune system damage, and individuals may show signs of early immune failure.

The state of viral infection in HIV-infected individuals is monitored by determining how actively the virus is replicating. Previously, the levels of infection were determined by testing for the amount of viral protein (the capsid protein or p24 antigen) in the blood. However this test was not very sensitive—during the asymptomatic period, p24 protein can generally not be detected, but it rises and can be detected as patients progress to AIDS. Currently the much more sensitive PCR-based viral load assay is used to measure the levels of circulating virus as described in Chapter 4. High viral loads are detected immediately after infection, and also as individuals progress to AIDS. As will be discussed, viral loads will climb in individuals on antiretroviral therapy if drug-resistant viruses develop. Therefore viral loads are monitored regularly in infected individuals receiving antiretroviral therapy.

Because development of drug-resistant virus is very important in treatment of HIV-infected individuals, detecting the presence of viruses that are resistant to particular antiretroviral drugs is important. Two types of tests are used. In *viral genotyping,* the PCR technique is used to amplify the viral RNA in the blood into viral DNA, and the sequence of the viral DNA is determined in specific regions. When HIV mutates to become resistant to an antiviral drug (e.g., AZT), characteristic alterations in the viral RNA sequence change the protein sequence of the targeted viral protein (e.g., reverse transcriptase). If the viral DNA sequence from the infected individual shows mutations characteristic of resistance to a particular antiretroviral drug, then it is very likely that the virus will be resistant to that drug. The second test for the presence of drug-resistant HIV is *viral phenotyping.* This is a more definitive, but laborious test.

Viral DNA is PCR-amplified from the blood of the infected individual, and key regions of the viral DNA are inserted by recombinant DNA techniques into a DNA copy of standard HIV. The resulting DNA is used to generate infectious HIV, and the resulting virus is then tested in the laboratory for resistance to different drugs. The knowledge of the drug resistance profile of HIV in an infected individual allows the doctor to select antiretroviral drugs that should control the infection.

Antiretroviral Therapy

Earlier in this chapter and in Chapter 4, we have seen how different anti-HIV drugs work. As discussed, treatment with individual drugs (monotherapy) has limited effectiveness due to the development of drug-resistant virus. The major breakthrough in management of HIV infection was the development of different classes of anti-HIV drugs. By combining anti-HIV drugs from these different classes, long-lasting control of infection has become possible. These highly effective combinations became available in 1997, when the first protease inhibitors were developed. This allowed combined treatment of patients with reverse transcriptase inhibitors and protease inhibitors.

The use of drug combinations in the treatment of many human diseases is well established. The treatment of cancer, another long-term disease, is a good example. Chemotherapy (drug treatment) for cancer became effective only after the development of drug combinations. At the maximum tolerable doses of one chemotherapy drug, most of the cancer cells in a patient may be killed, but a noticeable fraction will survive—typically 1 in 100 or 1 in 1,000. Those surviving cells will grow back and lead to recurrence of the tumor. However, if a second chemotherapeutic drug is combined with the first drug, the second drug can cause further killing of the tumor cells that survive the first drug. In practice, by the time three or four drugs are combined in a successful chemotherapy cocktail, they can lead to complete eradication of the tumor cells in the cancer patient.

Combination antiviral therapies for HIV-infected individuals are the standard approach to HIV treatment. Currently, the most effective and commonly used combinations involve two nucleoside analogs and one protease inhibitor or an NNRTI. For example, one combination is the nucleoside analogs AZT and 3TC along with the protease inhibitor indinavir. These combinations used to be referred to as "HAART" (highly active antiretroviral therapy, to distinguish them from the less effective single-drug monotherapies), but they are now simply referred to as ART (antiretroviral therapy). With effective ART therapies HIV-infected people experience a drop in viral loads in the blood to below the level of detectability, and this effect is sustained for several years. In general, the drops in viral loads in the blood are accompanied by recoveries of T_{helper} cell counts, increases in immune function, and the disappearance of AIDS symptoms.

Limitations and Uncertainties in ART Therapies

Despite the great success of ART in controlling HIV infection, there are limitations or uncertainties about the current combination therapies.

1. *The triple combination therapies are not effective in all people.* Although they have remarkable benefits for many HIV-infected individuals, they are less effective for others. For some people, viral loads in the blood have not shown very large decreases or have shown only temporary decreases. The triple combination therapies generally are less effective for HIV-infected individuals who were infected before the advent of ART therapies and who have previously taken antiviral drugs as single agents or monotherapies (e.g., AZT or other nucleoside analogs). These individuals generally already harbor HIV that is resistant to the nucleoside analogs. As a result, in these people the triple combination therapies are, effectively, more like single therapy with a protease inhibitor; as mentioned previously, protease-resistant HIV rapidly appears in individuals taking protease inhibitors alone. The continued development of more classes of antiviral drugs (e.g., integrase inhibitors) allows the possibility that new drug combinations can be applied that will control viral resistance to the original combination of drugs used.

2. *Uncertainty about the duration of effectiveness.* The triple combination therapies are now the standards of care for HIV-infected persons. However, it is not clear if any one ART combination can suppress viral RNA loads indefinitely. In fact, in extended clinical trials, the percentages of infected people who show complete suppression of viral RNA loads from a particular ART combination decrease. HIV mutants that are resistant to all drugs in a triple combination therapy are now appearing in infected people. Therefore, more antiviral drugs in existing and new classes will be key for long-term treatment of HIV-infected individuals.

3. *Drug side effects.* As discussed earlier in this chapter, many of the HIV antivirals have side effects. This is a serious problem because HIV-infected individuals need lifelong ART to control the infection. As a result, even relatively minor side effects may become serious after prolonged treatment. Part of the challenge for doctors managing HIV-infected patients is to find new ART combinations, if they can no longer tolerate the side effects of a given combination. For protease inhibitors, prolonged treatment can result in an unusual (but not life-threatening) side effect: redistribution of body fat to unusual locations, such as the back of the neck. More serious side effects include heart disease and diabetes.

4. *Difficulty in maintaining treatment schedules.* Taking ART therapies can be a complex matter. In a particular ART combination, an individual may need

to take as many as 25 pills per day at very precise intervals. Patients must faithfully take the different drugs at the prescribed times to keep the proper levels of drugs in their systems. Moreover, some of the protease inhibitors have very specific requirements for administration. For instance, the protease inhibitor indinavir must be taken on an empty stomach. The protease inhibitor ritonavir seriously affects the metabolism of other drugs that are used in the therapy of AIDS patients. These requirements present a serious challenge to doctors who are treating AIDS patients, as well as to the patients themselves.

In the ART combination therapies, strict adherence to the treatment regimen is very important. Receiving doses of drugs that are not optimal can actually accelerate a patient's rate of developing drug-resistant HIV. For protease inhibitors, once HIV resistance to one inhibitor has developed, resistance to other protease inhibitors often occurs. Maintaining *treatment adherence* is a particular challenge for HIV-infected individuals who already have difficult access to health care, such as economically disadvantaged people or injection drug users. Drug manufacturers are developing improved formulations that reduce the number of times a day that a drug has to be taken or that allow more than one drug to be taken in a single pill. For instance, Trizivir is a single pill combination of the drugs abacavir, AZT, and lamivudine. In 2006, a further advance was made with the approval of Atripla, a combination pill for the drugs efavirenz (NNRTI), emtricitabine (NRTI), and tenofovir (NRTI). Atripla is taken only once daily.

5. *Cost.* The combination ART therapies are very expensive. As discussed earlier in this chapter, the cost of AZT treatment alone is about $3,500 per year for a patient. Each drug in a triple combination costs at least that much; the current annual cost of a typical combination therapy is about $15,000 per person. In addition, if complications such as diabetes or high blood cholesterol develop, medications must be taken to control those conditions as well. If we consider that there are more than 1,000,000 people in the United States who are living with HIV infection, it would take approximately $15 billion yearly to provide ART to all of them. This is clearly a serious challenge to our healthcare system (this issue is discussed from the societal perspective in Chapter 11). We should also consider that worldwide, most of the cases of HIV infection and AIDS are in developing countries (see Chapter 6). These countries currently do not have the financial resources to provide ART to their infected populations.

Initially, introduction of ART led to a striking decrease both in the rates of AIDS deaths and also in the appearance of new AIDS cases (see Figures 6-1 and 12-2 and Table 12-2). However, for the past 5 years, the rates of death from AIDS have leveled off, and they are not continuing to decline. Thus, people with HIV/AIDS on ART are continuing to die, although at a slower rate. This trend appears to reflect the development of drug-resistant virus, drug side effects, and difficulties in treatment adherence. Clearly, it remains as important as ever to develop new and better anti-HIV drugs and also to develop effective measures to prevent new infections.

Timing the Initiation of ART

An important issue in management of HIV-infected individuals is when to initiate ART therapy. Arguments in favor of initiating therapy as soon as possible center on the desirability of reducing viral loads, which may slow progression of immunological defects. This may be particularly effective if the person is diagnosed during the initial period of acute infection when viral loads are very high. The possibility that ART may reduce the viral set point is important, since viral set points are correlated with the rate of progression to disease. Arguments in favor of delaying initiation of ART center both on the need to continue ART lifelong once it is initiated and on the possibility that drug-resistant virus may develop. As a result, beginning ART immediately after infection diagnosis heightens the risk of drug side effects since there will be prolonged treatment. Also, if drug-resistant virus develops, then that particular ART combination will no longer be available.

As a result, current recommendations are to monitor an HIV-infected individual and to initiate ART once the CD4 T_{helper} counts drop below 200 per cubic millimeter of blood. Initiation of ART is also recommended for individuals who have CD4 counts below 350 as well as an AIDS-defining illness. Initiating ART may also be considered for HIV-infected individuals if they are in the initial stage of viral infection (high viral loads, no antibodies yet); as mentioned, ART might reduce the viral set point in these individuals and thus reduce the likely rate of progression to disease. Prior to initiating ART, viral genotyping should be carried out to assist in selection of drug combinations that are likely to be effective in the infected individual.

Once infected individuals are receiving ART, they are regularly monitored for viral loads and levels of CD4 cells in the blood. In successful ART, the viral loads should be undetectable (fewer than 50 copies of viral RNA per cubic millimeter of blood), and the CD4 levels should be stable. An increase in viral loads and a drop in CD4 levels is an indication that the ART is no longer effective in that person due to development of resistant virus. Genotyping or phenotyping of the virus in such individuals will guide selection of a new ART combination that is likely to be effective (i.e., not resistant to the new drugs).

ART is also employed in the preventative (*prophylactic*) mode in limited circumstances. If healthcare workers become exposed to HIV (e.g., a needlestick with contaminated blood), it is recommended that they undergo ART for a fixed period of time to prevent the possible establishment of the viral infection.

Prophylaxis for Opportunistic Infections

HIV-infected individuals with full-blown AIDS (CD4 counts below 200) are at substantial risk for developing AIDS-related opportunistic infections. Therefore these individuals may be treated prophylactically with antibiotics to prevent occurrence of the corresponding opportunistic infection. *Pneumocystis* pneumonia is a common opportunistic infection in AIDS patients, so many are treated with preventative doses of trimethoprim and sulfamethoxazole (Bactrim) before they develop this complication.

---

http://biology.jbpub.com/fan/aids/6e/

Connect to this book's website: http://biology.jbpub.com/fan/aids/6e/. The site features summaries of the main points from each chapter, links to important AIDS-related websites, and short-answer-style review questions for each chapter.

CHAPTER 6
Epidemiology and AIDS

In Chapter 5, we considered how HIV manifests itself in an infected individual. The next level of complexity is to consider how HIV moves between individuals and its effects on populations. For these topics, the discipline of epidemiology is very important. This chapter gives an overview of epidemiology, with some applications regarding HIV and AIDS. The modes of HIV transmission and relative risk factors are addressed in Chapter 7.

Epidemiology is the study of the patterns of disease occurrence in populations and of the factors affecting them. This field is of great importance to the understanding of human diseases, and epidemiological studies can be used to address many questions. Epidemiological studies can:

- Identify new diseases
- Identify populations at risk for a disease
- Identify possible causative agents of a disease
- Identify factors or behaviors that increase risk of a disease and also determine the relative importance of a factor in contributing to a disease
- Rule out factors or behaviors as contributing to a disease
- Evaluate therapies for a disease
- Guide the development of effective public health measures and preventive strategies

It is important to keep in mind that epidemiological studies involve large groups, or *populations*, of individuals. This approach gives great power to these studies, because they draw on the total experience and behavior of large numbers of individuals.

Because epidemiological studies are based on observation of groups, some limitations and risks in interpretations are introduced. One limitation is that these studies cannot predict how any single individual will be affected by a factor even if the population as a whole is affected by that factor. Epidemiological studies also cannot predict the course a disease will take in a particular person.

Some risks are associated with drawing improper conclusions from epidemiology. For instance, it is important to avoid making an ecological fallacy: explaining behavior of an individual based on observations of an entire group. Another example of an improper conclusion is identifying certain characteristics of a group as causing a disease. For example, epidemiological studies have identified male homosexuals as one of the groups at high risk for AIDS. This does not imply that simply being homosexual causes AIDS, as some people have claimed. Instead, certain sexual behaviors by some homosexual men link them to AIDS, as we shall see later in this chapter and also in Chapter 7.

Despite these limitations and risks, epidemiological studies provide some of the most definitive information about the causes and dynamics of human diseases, short of carrying out experiments on humans.

An Overview of Epidemiology and AIDS

Epidemiology has played a central role in the fight against AIDS right from the beginning, and this will continue. The initial identification of AIDS as a new syndrome in 1981 was made through epidemiological studies. These studies reported the unusually high occurrence of individuals with rare diseases associated with immunological defects (see Chapter 1). The initial epidemiological studies showed a high frequency of the new disease in sexually active male homosexuals. Furthermore, the pattern of occurrence suggested that AIDS might be caused by an infectious agent that could be transmitted by sexual means. Subsequently, the appearance of AIDS cases among recipients of blood transfusions or blood products (for instance, hemophiliacs) and injection drug users suggested that AIDS could be transmitted through contaminated blood. The study of individuals afflicted with AIDS and also of groups of high-risk individuals led to the isolation in 1984 of HIV, the virus that causes AIDS. As soon as HIV was isolated, the virus was used to develop the test for HIV antibodies (see Chapter 4, p. 58).

The availability of the HIV antibody test allowed much more accurate epidemiological studies because evidence of infection could also be detected in healthy asymptomatic individuals. This led to the realization that an alarming number of individuals have been infected with HIV in many parts of the world. Moreover, we are seeing only the tip of the AIDS iceberg, because it often takes several years for the disease to develop. Epidemiological studies of high-risk groups have identified the underlying high-risk behaviors, such as unprotected sexual intercourse and sharing intravenous needles. This in turn has led to development of public health measures and safe-sex guidelines, which are our only weapons in AIDS prevention today. Finally, epidemiological studies (which also could be classified as clinical studies) provided the proof that AZT is an effective therapeutic drug for AIDS.

Basic Concepts in Epidemiology

The two basic kinds of epidemiological studies are *descriptive* and *analytical*. The goal of descriptive studies is to describe the occurrence of disease in populations. Analytical studies seek to identify and explain the causes of diseases. Frequently, descriptive epidemiological studies lead to analytical studies. For instance, descriptive epidemiology may identify a new disease, such as AIDS, or suggest hypotheses about the causes of a disease. Analytical studies will test the hypotheses and examine the disease in more detail.

Because epidemiology is the study of disease in populations, the proportion of affected individuals in a population is of basic importance. There are two important measures used in epidemiology:

Prevalence. This is the fraction (or proportion) of current living individuals in a population who have a disease or infection at a particular time.

Incidence. This is the proportion of a population that develops new cases of a disease or infection during a particular time period.

As an example, Table 6-1 shows the number of individuals with evidence of previous infection with hepatitis B virus (antibodies for the virus) in a city for the years 1968 and 1988. During this time, the population of the city has increased from 100,000 to 150,000. The *prevalence* of hepatitis B virus infection was 0.5% in 1968, and it increased to 0.67%

in 1988. The *incidence* of infection during this period was 0.17% (0.67% minus 0.5%); put another way, there were 170 new cases of infection per 100,000 people during the 20-year period. The yearly incidence rate would be 0.17% divided by 20 (0.0085%, or 8.5 new cases per 100,000 people per year). Epidemiologists use these prevalence and incidence data to calculate other expressions of their results, such as risk values.

Descriptive Studies

Descriptive epidemiological studies measure the appearance of disease by categories of person, place, and time. An example of disease appearance by *person* is the observation that lung cancer predominantly appears in individuals who smoke cigarettes (person = smokers). Disease appearance by *place*, for example, would show the low incidence of tooth decay in areas where there is a high level of naturally occurring fluorides in the water supply (place = high-fluoride areas). Disease appearance by *time* might be an outbreak of food poisoning resulting from contaminated food at a picnic (time = days after the picnic).

An important concept in descriptive epidemiology is clustering. *Clustering* is the unusually high incidence or prevalence of a disease in a subpopulation. Clustering can occur by person, place, time, or a combination. The first documented outbreak of Legionnaire disease is a good example of clustering. Legionnaire disease is a serious bacterial respiratory infection that can be fatal if untreated. The disease was first identified among several members of the American Legion who attended an American Legion convention at a hotel in Philadelphia in the summer of 1976. Thus, the disease was clustered with respect to place (the hotel in Philadelphia), time (1976), and person (American Legion members). Ultimately, a new microorganism (*Legionella*) was isolated that causes Legionnaire disease.

Types of Descriptive Epidemiological Studies

Descriptive epidemiological studies are carried out according to several design strategies or a combination of these strategies. Two of the important strategies are *case reports/case report series* and *cross-sectional/prevalence studies*.

Table 6-1	Hepatitis B Virus Infection in a City	
	1968	1988
Total population	100,000	150,000
Individuals with hepatitis B virus antibodies (seropositives)	500	1,000
Prevalence of seropositive individuals	0.5% (500/100,000)	0.67% (1,000/150,000)

■ Case Reports/Case Report Series

Case reports are descriptions of an unusual disease occurrence in individual patients. Sometimes the nature of the case may also suggest a relationship between some predisposing factor and the disease, or it may suggest the appearance of a new disease. These suggestions are strengthened if several similar cases are observed and reported together—a case report series. The original report in 1981 by Gottlieb describing *Pneumocystis* pneumonia in six homosexual men is a classic example of a case report series (see Chapter 1). This report suggested that a new disease (AIDS) might be occurring and that male homosexuals were at high risk.

■ Cross-Sectional/Prevalence Studies

In these studies, a population is monitored for the occurrence of a disease or a series of diseases, and statistics about each case (nature of the patient, geographical location) are recorded. This information can be used to construct a cross-sectional profile for the disease or diseases within the population. These studies are also sometimes carried out over a long period of time, and the date of disease occurrence is also recorded. Once the cross-sectional profile is obtained, it can be examined for clustering of disease cases by person, place, or time. These clusterings can suggest causes of known diseases and also identify new ones.

Cancer registries are an example of these studies. In these registries, information is gathered about all cases of cancer occurring in a region. Information from the cancer registry can then be used by cancer epidemiologists to investigate potential causes of cancer. For instance, these registries have provided strong evidence for a causal relationship between cigarette smoking and lung cancer.

The U.S. Public Health Service Centers for Disease Control and Prevention (CDC) maintains a registry of deaths and diseases, which is reported on a weekly basis in the journal *Morbidity and Mortality Weekly Report*. Information from this registry was also important in characterizing the beginning of the AIDS epidemic in 1981 and 1982 because there was a sharp increase in cases of *Pneumocystis* pneumonia and Kaposi's sarcoma at that time. The CDC now publishes a regular HIV/AIDS surveillance report that provides current and past epidemiological information exclusively on HIV infection and AIDS (see Appendix).

Prevalence studies can also be used to identify groups within the population that are at higher risk for a particular disease. In addition to suggesting possible causes of the disease, this information can be used for other purposes. First, if the disease is rare in the overall population, it will be more efficient to study the disease by focusing on the high-risk population; this is important for analytical epidemiology (see the following discussion). Second, public health workers may want to focus particular attention on the high-risk population as a first step in developing prevention strategies to combat the disease.

Analytical Studies

Analytical epidemiology studies are generally more focused than are descriptive studies. They investigate the causes of a particular disease, and they often involve assigning a numerical value to (*quantifying*) a potential risk factor. In fact, the distinction between descriptive and analytical studies is not absolute. Most epidemiological studies fall somewhere between a completely descriptive study and a purely analytical one. For instance, cancer registries can be used for analytical epidemiology studies, in which the relationship between a particular factor and a disease (e.g., smoking and lung cancer) is examined in detail, and the relative risk is determined.

Types of Analytical Epidemiological Studies

The two main approaches to analytical epidemiology are *experimental/ interventional studies* and *observational studies*.

Experimental/Interventional Studies

In these studies, a condition of an experimental subpopulation is changed, and the effect on the development of a disease is observed. The results are compared with the main population or an untreated subpopulation. This approach has been very useful in testing potential therapies for diseases. For instance, the Salk polio vaccine was tested in a nationwide trial of second- and third-grade school children in 1953 and 1954. The success of the trial led to the acceptance of the vaccine and the elimination of polio as a major health threat. A more recent example is a test of a vaccine for hepatitis B virus. This vaccine was tested in a population of sexually active male homosexuals (who are also at high risk for hepatitis B infection) and was shown to be very effective. Clinical drug trials can also be considered interventional epidemiology.

Although interventional studies are very useful in testing therapies, it is often difficult to use them to directly test if a factor causes a disease. Treating a group of people with a factor that might cause a disease raises serious ethical questions. This is particularly important if, as is the case for AIDS, there is no cure for the disease. One possible solution to this dilemma is to see if the same disease can be induced by the factor in animals. Another approach is observational epidemiology.

Observational Studies

Observational studies take advantage of the fact that within a population, some individuals will encounter a factor and develop a disease but others will not. The epidemiologist does not change the conditions of people to study the disease but rather *subdivides* the population according to possible risk factors for disease and studies them separately. For instance, by subdividing a population into cigarette smokers and nonsmokers, the epidemiologist can investigate the effect of cigarette smoking on cancer or heart disease without making anybody smoke. In some cases, a properly designed observational study can provide the same information as an interventional study.

The two main types of observational studies are: *case/control studies* and *cohort studies*. Case/control studies involve studying a group of individuals with a particular disease (the cases) and comparing them with a group of unaffected individuals (the controls). The controls are often matched for a number of factors not believed to be involved in the disease. If the cases differ from the controls by another factor as well, this would suggest that the factor is related to the disease. For instance, if lung cancer patients are compared with individuals without lung cancer, a higher percentage of the cancer patients are cigarette smokers than in the control population. Case/control studies are particularly useful if the disease being studied occurs only rarely. For instance, if a disease occurs only once per million people in the United States, it would be impractical to survey everybody in the population to study those few cases that occur. On the other hand, nationally, there would be about 300 cases of the disease, which could be readily studied by the case/control approach. For the same reason, case/control studies are important at the beginning of infectious disease epidemics, when there are still very few cases. The early epidemiology studies in the AIDS epidemic were mainly case/control studies.

Cohort studies focus on a group of individuals who share a particular risk factor for a disease. This group is then examined for the frequency or rate of disease appearance in comparison with a control population that does not have the risk factor. Such studies can implicate or exonerate a potential risk factor for the disease, and they can also determine the degree to which the risk factor contributes to the disease. Cohort studies can go forward or backward in time. *Prospective* cohort studies go forward in time, starting with an identified cohort of individuals and documenting development of disease as time progresses. A number of cohort studies for AIDS have been carried out, principally involving homosexual or bisexual men. For instance, one cohort study involved HIV antibody-positive individuals and tracked occurrence of lymphadenopathy syndrome (LAS), AIDS-related complex (ARC), and AIDS. In another study, a group of single men in San Francisco was followed for infection with HIV, and the factors (e.g., sexual practices, injection drug use) associated with infection were studied. Cohort studies that go back in time are called *retrospective* studies. In these studies, exposure to the risk factor has occurred previously, and the cohort of individuals is later identified for observation. For instance, retrospective cohort studies have been carried out on individuals who worked in asbestos-processing plants in the 1940s and 1950s. These individuals subsequently showed a high incidence of lung cancer, implicating asbestos as another potential cause of lung cancer.

Correlations

In analytical epidemiology, results are considered in terms of statistical associations, or *correlations*, between a factor and a disease. For instance, a high frequency of cigarette smoking is found among lung cancer patients, which means there is a statistical association between cigarette smoking and lung cancer. The aim of analytical epidemiology is to deduce *causality* from the statistical association. In our example, we would like to conclude that cigarette smoking causes lung cancer. However, statistical associations have other possible explanations.

Three alternate reasons for a positive correlation between a factor and a disease are as follows:

1. *There is no causal relationship.* This could result from faulty design of the experiment. For instance, if the control population is not properly matched with the experimental population, a false correlation could be observed.

2. *There is an indirect relationship.* In some situations a third confounding variable may influence both the factor being tested and the disease. For instance, there is a positive statistical correlation between alcohol consumption and lung cancer, but this does not mean that alcohol causes lung cancer. In this case, cigarette smoking is a confounding variable. Cigarette smoking causes lung cancer, and cigarette smoking is statistically associated with alcohol consumption. That is, individuals who are cigarette smokers also tend to drink more alcohol than do nonsmokers.

   Another example of an indirect relationship is the high statistical correlation between swimsuit sales and ice cream sales. This does not mean that ice cream consumption leads to swimsuit purchases or vice versa. In this case, summer or high temperature is a confounding variable. More swimsuits are bought during the summer when it is warm and beach weather is good, and more people eat ice cream during this time because the weather is hot.

3. *There is a direct causal relationship.* That is, a change in the factor will lead to a change in occurrence of the disease. It is important to remember that more than one factor can be a direct cause of a disease. Thus, establishment of a direct causal relationship between one factor and a disease does not rule out other factors as well.

Criteria for a Causal Relationship

In observational studies, it is difficult to absolutely prove a causal relationship from a correlation because the epidemiologist does not change the factors under study. However, certain criteria provide tests for causality:

1. *Strength of the association between the factor and the disease.* The strongest correlation would be if everybody with the factor gets the disease and nobody without the factor gets the disease. A strong correlation makes a causal relationship more likely. The argument is also strengthened if there is a dose–response relationship—that is, if individuals who have received higher exposure to a factor show higher frequencies of disease. However, it

is always important to keep in mind that confounding variables could exist.

2. *The association is consistent*—that is, the same correlation is observed in other studies, using different settings and different populations.
3. *The association has the correct time relationship*—that is, exposure to the agent must occur before development of the disease.
4. *The association has biological plausibility*—that is, association of the factor with the disease makes biological sense.

For infectious agents, another set of rules has been developed for assessing whether a microbe causes a disease: *Koch's postulates*. Koch's postulates are discussed in Chapter 2, p. 13, and they require both observational and experimental studies. Briefly, a microorganism can be considered the cause of a disease if (1) it is always found in diseased individuals, (2) it can be isolated from a diseased individual and grown pure in culture, (3) the pure microorganism can cause the disease when introduced into susceptible individuals, and (4) the same microorganism can be reisolated from those individuals.

Epidemiology and AIDS in the United States

Let us now see in more detail what epidemiology can tell us about AIDS. As described in the overview to this chapter, epidemiology has been extremely important in this epidemic. Let us look at some of the epidemiological information and the conclusions that can be drawn from it.

The Current Picture of AIDS in the United States

Figure 6-1 shows the total number of AIDS cases that have been reported in the United States for the years 1978 through 2007. By the end of 2007, a total of 1,018,875 individuals had reportedly developed AIDS, of which 583,298 had died. Current estimates are that approximately 1.2 million people in the United States are infected with HIV; many of these people may develop AIDS and ultimately die if they do not have access to ART or if the therapies are not completely effective. Thus, we can see the seriousness of this epidemic and the strain that it places on our society.

Figure 6-2 shows the distribution of AIDS cases according to risk groups. Homosexual and bisexual men make up the largest percentage of cases, followed by injection drug users, heterosexual partners of HIV-infected individuals, and children of HIV-infected mothers. In the early part of the AIDS epidemic, hemophiliacs and recipients of blood or blood products also were significant risk groups; however, the HIV ELISA test has largely eliminated HIV from the blood supply, and these individuals now make up less than 1% of cumulative AIDS cases. For a small percentage of cases (about 1%), no risk group has been assigned, perhaps because of the unavailability of information about the patients or their reluctance

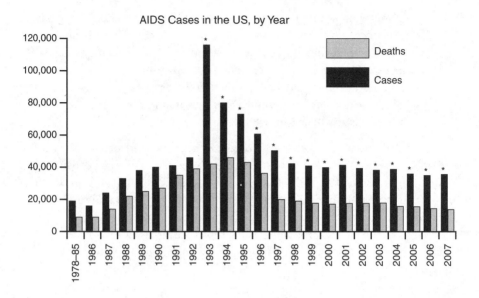

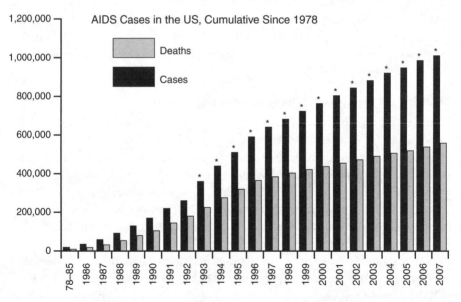

Figure 6-1 Appearance of AIDS in the United States. Data compiled from reports from the U.S. Centers for Disease Control and Prevention. In 1993, the definition for AIDS was expanded to include any HIV-positive person with a T_{helper} count of less than 200 per cubic millimeter of blood. This accounts for the jump in the number of reported AIDS cases beginning in 1993. Data for all subsequent years use this new definition (designated by * on the graphs). (Data from CDC, *HIV/AIDS Surveillance Reports.*)

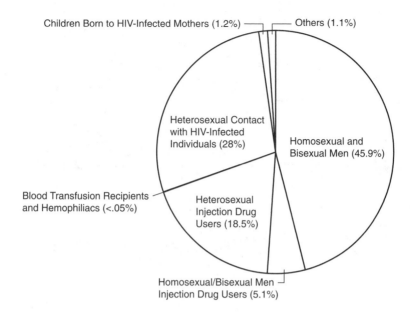

Children Born to HIV-Infected Mothers (1.2%) ── ┌── Others (1.1%)

Heterosexual Contact
with HIV-Infected
Individuals (28%)

Homosexual and
Bisexual Men (45.9%)

Blood Transfusion Recipients
and Hemophiliacs (<.05%)

Heterosexual
Injection Drug
Users (18.5%)

Homosexual/Bisexual Men ┘
Injection Drug Users (5.1%)

Figure 6-2   Distribution of AIDS cases in the United States by risk group. (Cumulative figures, 1981 to December 2007; based on data from states where results from confidential HIV testing are reported.) (Data from CDC, *HIV/AIDS Surveillance Reports.*)

to acknowledge membership in a high-risk group. From the beginning of the AIDS epidemic women make up 19% of American AIDS cases; this low percentage is because the largest number of cases occur in homosexual and bisexual men. However, they make up 27% of HIV-infected people living in the United States today.

Figure 6-3 shows the distribution of AIDS cases according to ethnicity. There is a disproportional number of AIDS cases among minorities, particularly African Americans and Hispanics. Indeed, although these groups make up about 21% of the general population, they make up 59% of AIDS cases—and an even higher percentage (80%) of the cases associated with injection drug use. Put another way, the frequency of AIDS cases among African Americans and Hispanics is about three to five times higher than in the general population. In fact, AIDS is currently the leading cause of death of African American women 25–34 years of age and of African American men 35–44 years of age. This points out the urgency of developing public health and educational measures targeted to these communities to control the epidemic.

The epidemiological data described in this section provide a current picture of HIV/AIDS in the United States. It is important to bear in mind that this picture can change as the AIDS epidemic progresses (as discussed on p. 112).

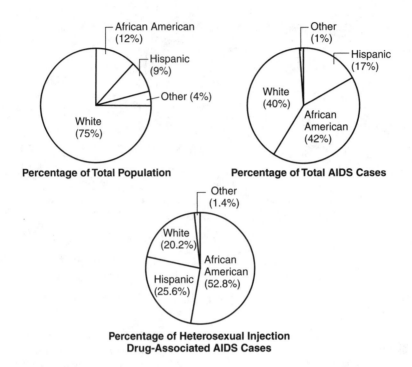

Figure 6-3 AIDS cases by ethnicity in the United States. (Cumulative figures, 1981 to December 2007.) (Data from CDC, *HIV/AIDS Surveillance Reports.*)

Epidemiology and Modes of HIV Transmission

Transmission of HIV is addressed in detail in Chapter 7. However, some examples of epidemiological studies regarding HIV transmission are presented here to illustrate how these studies allow us to draw conclusions about the relative risks of different activities for HIV transmission. One study implicates an activity (anal sex) in HIV transmission, and another study shows that casual contact does not cause HIV transmission. In addition, the possibility of HIV transmission by insects is considered.

Anal Sex—A High-Risk Mode

Let us consider an epidemiological study that looked at the relative risks of different sexual activities. This study was part of the San Francisco Men's Health Study, which is a cohort study of single men in an area of that city. This particular area has been hit especially hard by the AIDS epidemic. The study involved 1,034 single men, who were monitored for HIV antibody status and asked about their sexual practices. Of the homosexual men in this cohort, 48% were seropositive at the beginning. A low percentage (17.6%) of men who had refrained from sex during the previous 2 years were seropositive, and this could be traced to sexual activity before that time. Table 6-2 shows frequencies of HIV infection when the homosexual men in the study were divided according to whether or not they practiced unprotected anal intercourse. Those who were the receptive partner or who were both receptive and insertive showed significantly higher frequencies of HIV infection than those who did not engage in anal sex. This finding shows that unprotected anal intercourse is a high-risk mode of HIV infection.

The results also showed that those who practiced only insertive anal intercourse were at less risk; in fact, this particular study could not statistically distinguish those men from men who did not engage in anal sex at all. However, other studies of heterosexual couples (who engage in vaginal as well as anal intercourse) clearly show that insertive intercourse can result in HIV transmission to the man. Thus, the most likely conclusion is that unprotected anal receptive intercourse is a very high-risk sexual activity, and insertive intercourse is somewhat lower, although significant, in risk.

Casual Contact—No Measurable Risk for HIV Transmission

Casual contact with HIV-infected individuals poses no risk for infection. This was determined early in the epidemic because people living with AIDS patients did not develop signs of HIV infection or AIDS. Casual contact includes hugging, touching, dry kissing, sharing of eating or drinking utensils, and sharing the workplace, telephones, and the like.

An example of an epidemiological study establishing that casual contact does not lead to HIV infection is shown in Table 6-3. One hundred one individuals who shared a household with an AIDS patient for at least 3 months were tested for the presence of HIV antibodies. Only one person was seropositive, and this individual was a child

Table 6-2	HIV Infection in Homosexual Men: The Relative Risk of Anal Sex*
Sexual Practices for the Preceding 2 Years	Percent HIV Seropositive (Adjusted for Number of Sexual Contacts)
No anal sex	20.6%
Unprotected anal sex, insertive only	26.7%
Unprotected anal sex, receptive only	44.6%
Unprotected anal sex, both insertive and receptive	53.3%

*Data from the San Francisco Men's Health Study by W. Winkelstein, et al., *J. Am. Med. Assoc.* 257 (1987): 321-325. Individuals who did not practice anal sex for the previous 2 years included those who practiced oral sex only and also those who abstained entirely. The strongest correlation for HIV seropositivity in this study was the number of sexual contacts an individual had. The percentages in the table were adjusted to account for the average number of sexual contacts for the different groups.

of two injection drug users. Further investigation showed that this child acquired the infection at birth and not through casual contact. Thus, none of the individuals in this study became HIV infected through casual contact.

The results in Table 6-3 show that the risk of contracting HIV by casual contact is quite low, but this study by itself could not rule out the possibility that casual contact could result in infrequent spread of HIV infection (because the number of subjects studied was not large enough). However, in the more than 20 years that have elapsed since this study was reported, no documented cases have been reported of HIV transmission by casual contact. Thus, it is safe to state that casual contact does not lead to transmission of HIV.

Insect Bites—No Evidence for Spread of HIV

Early in the AIDS epidemic, some people worried that insect bites could be a source of infection because insects such as mosquitoes draw a blood meal from the person they are biting, and they move from person to person. Some other viruses (e.g., yellow fever) are indeed transmitted by mosquitos. However, epidemiological evidence argues against insect transmission of HIV. First, in well-studied North American or European populations, the great majority (almost 99%) of AIDS cases can be explained by the well-documented modes of transmission. This includes a study in Belle Glade, Florida, where some people proposed an outbreak of AIDS due to insect transmission.

Table 6-3	HIV Infection in Casual Household Contacts of AIDS Patients*	
	Number Tested	Number HIV Seropositive
Children less than 6 years old	21	1†
Offspring of an AIDS patient	15	1†
Offspring of others	6	0
Children 6 to 18 years old	47	0
Adults	33	0
Total tested	101	

*The subjects in this study had lived in the same household with an AIDS patient for at least 3 months. Individuals who were in known high-risk groups (sexual relations with the AIDS patient, injection drug use, homosexual men) were not included. Thus, only individuals who had casual contact with the AIDS patient were studied. Among this group, 48% shared drinking glasses with the AIDS patient, 25% shared eating utensils, 9% shared razor blades, 90% shared toilets, and 37% shared beds. The AIDS patient was hugged by 79% of the subjects, kissed on the cheek by 83%, and kissed on the lips by 17%.

†Further investigation showed that the one seropositive child was the offspring of two injection drug abusers and probably acquired the infection at birth; this is a known mechanism of HIV transmission. Thus, none of these casual household contacts of AIDS patients became infected. Data from a report by G. H. Friedland, et al., *N. Engl. J. Med.* 314 (1986): 344–349.

Furthermore, in Africa, where the insect populations and prevalence of HIV infection are high, the age and geographical distributions of HIV infection argue against insect transmission. HIV is rare in children and the elderly, even in households where there are HIV-infected individuals. The old and the young are actually more frequently bitten by mosquitoes and other insects. In terms of geographical distribution, HIV infection is at high frequency in certain cities and urban areas, and there is much lower frequency of infection in surrounding rural areas. This is actually opposite from the pattern that would be expected if insects transmitted HIV because they are more plentiful in rural areas.

Likelihood of Progression to AIDS

One very important concern is the likelihood of an HIV-infected individual eventually developing clinical AIDS. Prospective cohort studies have followed HIV-seropositive individuals for development of AIDS or ARC. An example of one study is shown in Table 6-4. After about 4 years of observation, 18% of the seropositive individuals developed AIDS, and an additional 47% developed signs of immunological impairment. Only 35% remained asymptomatic. Other similar studies predict that in the United States, most individuals infected with HIV (i.e., more than 70%) will develop AIDS within 10 to 12 years of infection if they do not receive antiviral therapy.

The Effectiveness of AZT

As described in Chapter 4, p. 62, AZT was the first antiviral agent effective in AIDS. Previous laboratory experiments had shown that the drug could block HIV infection in isolated culture systems, and the drug was tested in a clinical trial, which can also

Table 6-4	Long-Term Results of HIV in Infected Men*		
		Number of Individuals	Percent Total
AIDS		10	17.5
Signs of immunological damage			
LAS		16	28.1
Others (oral candidiasis, weight loss, etc.)		11	19.3
Subtotal		27	47.4
Asymptomatic		20	35.1

This study was carried out before antiviral drugs were available. Today the development of AIDS would be much slower in individuals taking triple combination therapies.

*Men in this study were followed for an average of 44 months after they showed initial signs of HIV infection (seroconversion). Data from G. Wormser, et al. *AIDS: Acquired Immune Deficiency Syndrome and Other Manifestations of HIV Infection*. Noyes Publications, 1987.

be considered an interventional epidemiological study. Results from the first clinical trial of AZT are shown in Table 6-5. AIDS patients who had experienced one bout of *Pneumocystis* pneumonia were divided into two groups. One group received AZT, and the other control group received placebo pills that contained no drug. The physical states of all subjects were then monitored on a regular basis. After the study was in progress for less than 6 months, the results showed markedly better survival of the group taking AZT than the control group. In fact, the results were so striking that the investigators terminated the trial early and administered AZT to the control patients as well. Withholding the drug would have been unethical at that point. These studies led to approval of AZT for therapy in AIDS. Similar studies led to the approval of the other HIV antiviral drugs available today.

The Changing Face of AIDS

Earlier in this chapter we saw the epidemiological statistics for HIV and AIDS in the United States in cumulative terms (Figures 6-1, 6-2, and 6-3). It is important to remember that these statistics represent the cumulative experience with HIV infection since the beginning of the epidemic. However, the distribution of HIV infection in populations at risk changes with time; as HIV infection moves through different populations at different rates, the distribution of infection changes. Some examples are shown in Figures 6-4 and 6-5. From the beginning of the AIDS epidemic through 1986, homosexual and bisexual men made up 73% of the AIDS cases. By 2007, they represented 51% of the cumulative AIDS cases. This change represents the fact that relative to other risk groups, spread of HIV infection in homosexual and bisexual men during the 1980s and early 1990s increased more slowly. Likewise, in 1986, 60% of cumulative AIDS patients were non-Hispanic whites, but in 2007, only 40% were. This reflects the HIV/AIDS epidemic disproportionately affecting the Hispanic and African American communities in recent years. In 2007, 49% of new AIDS cases were among African Americans, 19% were among Hispanics, and 29% were among non-Hispanic whites.

Table 6-5	Effect of AZT Treatment on Survival of AIDS Patients*		
Treatment	Number of Subjects	Number of Deaths	Percent Deaths
None (Placebo pills)	97	19	19.6
AZT	124	1	0.8

*The subjects were in the study an average of 16–17 weeks. The study was intended to last 24 weeks (6 months), but it was terminated early once the dramatic effect of AZT treatment became evident. All subjects were offered AZT at that time. Data from the report of the first large-scale test of AZT by M. A. Fischl, et al., *N. Engl. J. Med.* 317 (1987): 185-191.

The changes in AIDS distribution also make clear that current statistics on HIV/AIDS distributions do not tell us where HIV infection is spreading most rapidly. For instance, in the United States, young adults are among the groups in which HIV infection is increasing most rapidly, although they do not currently make up the largest

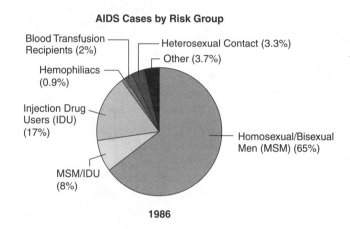

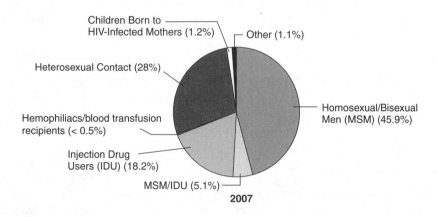

Figure 6-4 Cumulative AIDS cases by risk group are shown as of 1986 and as of 2007. These statistics represent the total cases since the beginning of the epidemic, and AIDS takes several years to develop. Therefore, the proportions of new HIV infections associated with homosexual/bisexual men may be even lower, and the proportions associated with heterosexual contact and injection drug use may be even higher. Homosexual and bisexual men are also referred to as men who have sex with men (MSM). (Data from CDC, *HIV/AIDS Surveillance Reports*.)

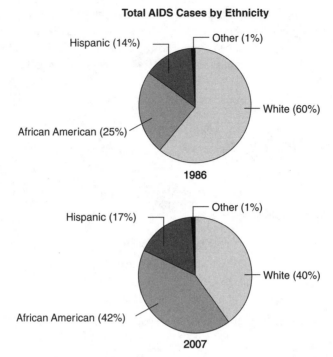

Figure 6-5 Cumulative AIDS cases by ethnicity are shown as of 1986 and as of 2007. These represent the total cases since the beginning of the epidemic to those two dates. (Data from CDC, *HIV/AIDS Surveillance Reports*.)

number of AIDS cases. Development of clinical AIDS takes several years, so individuals who become infected with HIV today may not impact the statistics for several years.

In the next section, we will see that the worldwide distribution of HIV infection and AIDS cases is also changing rapidly.

AIDS Around the World

AIDS was first recognized as a disease in the United States. However, from the worldwide perspective, HIV infections in the United States and western Europe represent a small fraction of the total cases. Estimates by public health officials indicate that in 2008, 33 million people in the world were living with HIV infection and/or AIDS. Approximately 94% of HIV-infected people are currently living in developing countries, with more than 82% living in sub-Saharan Africa or southern and Southeast Asia alone. The distribution of HIV infection throughout the world is illustrated in Figure 6-6.

It is important to consider control of HIV/AIDS in the global perspective for two reasons. First, HIV-infected people in all areas of the world develop the same immunodeficiency that is seen in developed countries, and the amount of suffering is equivalent. Second, it is in the self-interest of developed countries to combat infections such as HIV wherever they occur worldwide. In this age of rapid transportation, an infectious disease is only an airplane ride away from any place in the world. Thus, any HIV in the world is a threat to those of us in the United States. It is in our own self-interest to combat the spread of HIV on any continent.

AIDS in Africa

AIDS is a major health problem in sub-Saharan Africa. In 2008, it was estimated that there were 22 million HIV-infected individuals in Africa, many of whom will probably progress to AIDS and eventually die. The disease was initially centered in countries of central Africa, including the Democratic Republic of the Congo (Zaire), Kenya, Uganda, Zambia, Tanzania, and Rwanda. More recently, the prevalence of HIV infection in those countries, while high, has begun to level off or drop somewhat, due to (1) increased prevention efforts and (2) death of infected individuals. On the other hand, the epidemic is growing rapidly in southern Africa—South Africa, Zimbabwe, Botswana, and Swaziland. In contrast to the distribution of cases in North America and Europe, HIV infection is somewhat higher in women than in men; epidemiology shows that the predominant mode of transmission is heterosexual intercourse. The initial epidemic spread along truck routes through central Africa, and female prostitutes were important reservoirs for the infection. The AIDS epidemic is mostly concentrated in cities and

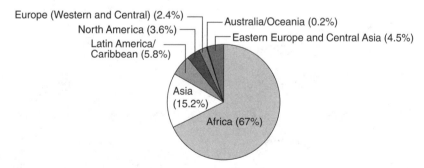

Figure 6-6   The worldwide distribution of HIV infection. The distribution shown is taken from estimates as of December 2007 by the UNAIDS 2008 report on the global AIDS epidemic. These are estimates because reporting of HIV infection is incomplete in some parts of the world. (Data from UNAIDS, *Report on the Global AIDS Epidemic [2008]*.)

urban areas and is lower in rural areas. As in North America and Europe, the spread of HIV infection in Africa is a recent phenomenon, beginning mostly in the 1980s.

In the past decade, the most rapid increases in HIV infection have been occurring in southern Africa. South Africa is the country with the most HIV-infected people in the world (estimated 5 million). The prevalence of HIV infection among young adults 15–24 years of age is 17% in females and 4.5% in males; approximately 30% of pregnant women attending prenatal care clinics are infected. Neighboring Swaziland has among the highest prevalences of HIV infection, with 23% of females and 6% of males ages 15–24 years infected; and 40% of pregnant women are infected. These very high rates of HIV infection will have devastating social and economic impacts on these countries that may reach the proportions of some of the ancient epidemics described in Chapter 2. In southern Africa the epidemic has resulted in substantial premature deaths, particularly among people of 30–50 years of age, who are normally the most productive in a population. Figure 6-7 shows the population distribution of people in Lesotho, a country within southern Africa with high HIV infection, in 1950 versus 2007. The relative numbers of people in the 30- to -50-year age group have declined, with increases in young children. This places a substantial strain on the society. In South Africa the life expectancy has actually declined from 61 years in 1990 to 49 years in 2007, and this is largely due to the AIDS epidemic.

The high prevalences of HIV infection among women in southern Africa has also resulted in relatively large numbers of HIV-infected children, who largely acquire infection during or shortly after birth. In 2007, there were 270,000 deaths

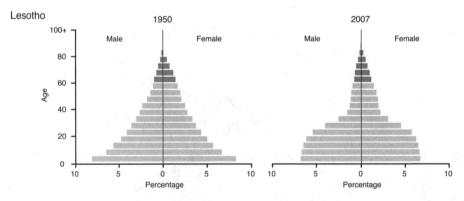

Figure 6-7 The population distribution according to age is shown for Lesotho, a country in southern Africa, for 1950 and 2007. The change in the population distribution reflects deaths from AIDS, which has particularly affected people 30–50 years of age. (Reproduced from UNAIDS, *Report on the Global AIDS Epidemic* (2008). Original source material from Population Division of the Department of Economics and Social Affairs of the United Nations Secretariat, World Population Prospects [2006].)

from AIDS in children worldwide, and more than 90% of them were from sub-Saharan Africa.

The magnitude of the HIV/AIDS epidemic in Africa is compounded because these developing countries have limited financial abilities to provide medical care to HIV-infected patients. In particular, antiviral drugs (most notably the ART combinations), as well as drugs for many of the opportunistic infections, are almost beyond the reach of most HIV-infected patients in Africa. A recent positive development is that AIDS drug manufacturers have reached agreements that allow these drugs to be manufactured and distributed in developing countries at substantially lower costs than charged in developed countries. Nevertheless, intervening in HIV transmission through preventive measures (and ultimately through the use of a vaccine) takes on even greater importance.

One factor that has increased the rates of HIV transmission in African populations is co-infection with sexually transmitted diseases that cause sores in the genital tract tissues (for instance, syphilis and chancroid). This relationship makes sense because these infections allow the HIV virus or HIV-infected cells easier access to CD4 T-cells during sexual intercourse. Epidemiological studies have found that public health measures that decrease sexually transmitted disease infections also decrease the rate of HIV transmission (Figure 6-8). Circumcision in males also is associated with lower

Figure 6-8   A woman walks past an AIDS awareness banner at Rand Afrikaans University in Johannesburg, South Africa, on World AIDS Day, December 1, 2002. (© Paul Botes/AP Photos)

rates of HIV infection, perhaps due to decreased frequencies of sores from the secondary sexually transmitted diseases.

As mentioned in Chapter 4, another virus related to HIV has also been discovered in Africa. The original HIV that is associated with the great majority of AIDS cases is called *HIV-1* (see Chapter 4, p. 52). The other virus is called *HIV-2* and is predominantly found in countries along the west African coast, such as Senegal and the Ivory Coast. Molecular biological experiments tell us that HIV-1 and HIV-2 are closely related but distinct viruses that evolved from a common ancestor thousands of years ago. HIV-2 also causes AIDS, although there are some indications that it is less able to cause disease than HIV-1.

The existence of HIV-2 raises a problem because the standard HIV-1 ELISA tests do not detect HIV-2 antibodies. Thus, the standard HIV test will not detect individuals infected with HIV-2. So far in North America, few cases of HIV-2 infection have been found. However, it may be important to screen blood supplies and individuals for HIV-2 infection as well to avoid undetected contamination or infections.

AIDS in Asia

While HIV infection first spread explosively in Africa, it reached Asia in the mid-1980s and began to spread rapidly. Currently, Asia is the area with the second largest number of cases of HIV, with an estimated 5 million in 2008. In Asia, HIV infection initially began to spread among commercial sex workers and injection drug users in Thailand, Myanmar (Burma), and Malaysia. In these countries, heterosexual sex and injection drug use are the major routes of infection. The levels of infection and the rates of increase in the high-risk populations have been well-documented and are striking. For instance, in one 6-month period (January to July 1988), the percentage of HIV-positive injection drug users monitored by one Bangkok hospital climbed from 1% to more than 30%. Fortunately, with the adoption of new public health measures (notably increased condom use) in the mid-1990s, along with introduction of ART therapies, the rate of new infections in Thailand began to level off. Unfortunately, recently the rates of HIV infection have begun to increase again in high-risk groups due to complacency and less frequent condom use or other prevention measures. The prevalence of HIV infection in male homosexuals in Thailand increased from 17% in 2003 to 28% in 2005. Currently, more than 10% of female prostitutes and 45% of injection drug users are HIV infected in Thailand. While the rates of infection have been decreasing or leveling off in Cambodia, Myanmar, and Thailand, they have been climbing in Indonesia, Pakistan, and Vietnam.

HIV infection reached India somewhat later than Southeast Asia. Although the overall prevalence of infection is fairly low, India is the second most populous country in the world, so the numbers of HIV-infected individuals are substantial. In India, the predominant modes of HIV infection are heterosexual sex and injection drug use. In some populations of female prostitutes and injection drug users, the prevalence of HIV infection is greater than 50%.

HIV infection has also moved into China, the world's most populous country. Approximately 650,000 people are infected. In China, the epidemic is predominantly among injection drug users, where sharing of needles is commonplace. Infection largely entered the country from Southeast Asia along drug supply routes. Approximately 50% of Chinese injection drug users are HIV infected.

New Areas of Rapid Spread

HIV infection has moved into several other geographical areas of the world. Repeating the patterns seen in other earlier affected areas, infection initially spreads rapidly and silently once it enters a high-risk population, before effective public health measures can be put in place. New epidemics are occurring in eastern and central Europe, where there has been a 20-fold increase in HIV infections during the past 10 years. HIV infection is rapidly growing in Russia and the Ukraine. Russia now has the largest HIV epidemic in Europe, with an estimated one million people infected. HIV infection is spreading even more rapidly in the Ukraine. The predominant risk factor is injection drug use. Recently, another region where HIV has been spreading rapidly is in the island of Papua New Guinea in the South Pacific.

It is important to note that these figures for HIV infection around the world represent HIV-infected individuals (detected by the antibody test described in Chapter 4, p. 58), not AIDS cases. Because many of these infected individuals acquired the virus relatively recently, most have not begun to show signs of illness yet, particularly in regions where the spread of HIV has been recent. However, we can predict that within a few years, the number of AIDS cases in these countries will soar, as has already happened in central and southern Africa, North America, and western Europe. If there are not very many cases of full-blown AIDS in a country yet, it is easy for the general public (and politicians) to ignore or downplay the problem for the time being.

HIV Subgroups and Clades

Different types of HIV-1 infect different parts of the world. These differences are largely in the envelope gene of the virus. HIV-1 has been divided into three subgroups: subgroup M (the major subgroup) and two minor *subgroups* (N and O). The HIV-1 in subgroup M has been further divided into subtypes, or *clades*; there are now more than 10 clades of HIV-1. In the United States and northern Europe, the predominant HIV-1 is clade B. In Africa, most of the HIV-1 clades are found, but the predominant virus is clade C. In Asia, the predominant infections are by clade C and clade E viruses. The existence of different HIV-1 clades has practical implications. First, distinguishing between different clades has allowed epidemiologists to more accurately track the spread of infection from one location to another. For instance, there are actually two HIV epidemics going on in Thailand: One epidemic, involving heterosexual transmission, is in the northeastern part of the country (clade C or E HIV-1), and the other epidemic, involving injection drug use, is centered around Bangkok (clade B HIV-1). Second, the envelope proteins of the different HIV-1 clades are sufficiently different that some ELISA tests do not detect viruses of very different clades, although this is currently less of a problem. It has also been suggested that HIVs of different clades may differ somewhat in their ability to establish infection and/or cause disease. Finally, it is possible that different clades of HIV-1 may not respond equally well to existing or future antivirals or vaccines. Prior to treating HIV infections in different parts of the world, it will be important to assess the effectiveness of these agents on the clade(s) of HIV infecting those areas.

The worldwide nature of HIV infection makes it a very important public health problem. No continents or countries are safe from infection, and the virus can spread rapidly (and undetected) once it enters a group engaging in high-risk behaviors.

http://biology.jbpub.com/fan/aids/6e/

Connect to this book's website: http://biology.jbpub.com/fan/aids/6e/. The site features summaries of the main points from each chapter, links to important AIDS-related websites, and short-answer-style review questions for each chapter.

CHAPTER 7

Modes of HIV Transmission and Personal Risk Factors

Biological Bases of HIV Transmission
Sources of Infectious HIV
Stability of HIV
Targets for HIV Infection

Modes of HIV Transmission
Activities Not Associated with HIV Transmission: Casual Contact
Activities Associated with HIV Transmission: Birth, Blood, and Sex

In previous chapters, we analyzed how the AIDS virus operates at the cellular level and at the organism level. Now our focus shifts to the interorganism level. In this chapter, we look specifically at the question of how HIV is transmitted from person to person. Because there currently is no cure for AIDS, once an individual has contracted the disease, preventing the transmission of HIV from person to person is critical. Consequently, in this chapter we consider risk factors for HIV transmission and discuss ways of reducing these risk factors.

The evidence for assigning risks to different levels of activities comes from two main sources: theoretical biological analysis and empirical epidemiological data, bolstered by laboratory data. Theoretical analysis considers the biological plausibility of HIV transmission for particular activities based on the presence or absence of substances containing HIV and of receptors for these substances. For example, we know that HIV is not present in someone's exhaled breath; consequently, on the basis of theoretical analysis alone, we would assign little risk to breathing the air in the same room with a person with AIDS. Theoretical analysis can be used to make predictions about no-, low-, or high-risk activities. These predictions ultimately can be tested by empirical epidemiological data, the other main source of evidence for our risk judgments in this chapter. To continue the example of transmission by breathing, epidemiological data from the sample of individuals who have lived with people with AIDS provide corroborating evidence that breathing the same air does not spread HIV (see Chapter 6, Table 6-3). Because epidemiological data indicate no AIDS incidence among family and friends who have simply lived with people with AIDS without intimate contact, and because of the biological implausibility, we can confidently state that breathing the same air is not a risk factor.

Typically, it is those activities with a high biological plausibility of HIV transmission that are carefully investigated with epidemiological studies. We saw one example of this in the last chapter. Anal receptive sexual activity has a high biological plausibility of HIV transmission, and the evidence from epidemiological studies discussed in Chapter 6 (Table 6-2) provides corroborative evidence that this behavior is, in fact, strongly associated with HIV infection.

In addition, epidemiological evidence can provide the initial evidence that certain activities are or are not associated with HIV infection risk. At the outset of the AIDS epidemic, for instance, it was epidemiological studies that led to the identification of likely modes of HIV transmission. This, in part, guided subsequent biological laboratory work, aided in the theoretical understanding of AIDS, and resulted in the isolation of HIV.

We should remember one aspect of epidemiological information at the outset of our discussion of risk and risk factors. Epidemiological studies, by their nature, identify groups of individuals with diseases, but it is wrong to conclude that there is something about the group itself that causes the disease. For instance, AIDS epidemiological studies have identified certain groups of individuals who are overrepresented in the population of those with AIDS, compared with their representation in the general population. These groups include homosexual and bisexual men and African American males. However, there is nothing about being homosexual or bisexual or an African American male that, by itself, leads to HIV infection and AIDS. Rather, some individuals in these groups are more likely to undertake certain *behaviors* that have a high biological plausibility of HIV transmission. Because of these higher HIV risk behaviors among some people in the group, the likelihood of transmission *averaged over the whole group* increases if the necessary and sufficient behaviors for HIV transmission occur. It is important to focus on the *behaviors* as the causal factor, not the group association.

Before we consider the risks of particular behaviors, however, we need to understand the biological bases of HIV transmission, including such issues as the primary sources of HIV within an infected person, the stability of the virus in moving between individuals, and the targets for infection in an uninfected individual.

Biological Bases of HIV Transmission

In infected people, infectious HIV is present only in cells and some human body fluids. Despite its devastating effects within the body, the virus is actually quite fragile in the external environment and dies quickly when exposed to room temperature and air conditions. In fact, very special laboratory conditions are needed to grow HIV outside the human body. It is important to remember this fact because it is easy to assume—mistakenly—that a disease as deadly as AIDS must be caused by an agent that is tremendously strong and sturdy. People's fears of the disease, combined with their lack of knowledge and mistaken impressions about epidemics, can cause them to view HIV in an anthropomorphic way—almost like a living, breathing enemy capable of thought and devastating action. The reality of HIV outside the body is much different: It is a fragile virus that loses infectivity quickly.

Sources of Infectious HIV

In an infected individual, HIV is present in particular cells and in some bodily fluids and secretions, many of which also contain these particular cells. In terms of cells, macrophages and T_{helper} lymphocytes are susceptible to infection by HIV, as described in Chapter 4. Macrophages may be the long-term reservoirs of HIV in infected individuals because they are not killed by the virus. Macrophages circulate through the bloodstream, and they also are found in all mucosal linings of the body, such as the internal urogenital surface of the vagina and penis and the lining of the anus, lungs, and throat. Another kind of cell that can be infected with HIV is the Langerhans cell, found on mucosal surfaces and below the surface of the skin.

Among people who test positive for HIV, the virus is not found consistently in all body fluids and products. Furthermore, in body fluids where HIV is regularly found, it occurs in different concentrations at different times. Nonetheless, we can place the body fluids and products into four groups based on the degree of association between body fluids/products and HIV infectivity. These groupings reflect differences among body fluids/products in their general concentrations of infectious HIV or HIV-infected cells and in the amount of relative exposure a typical individual might experience. Table 7-1 lists these groupings.

Table 7-1	Degree of HIV Infectivity of Different Body Fluids and Products of Individuals Known to Be HIV Positive
Group 1: *Very high infectivity**: Blood, semen, vaginal/cervical secretion (including menstrual fluid)	
Group 2: *High infectivity*: Breast milk	
Group 3: *Low infectivity*: Saliva, tears	
Group 4: *No infectivity*: Perspiration/sweat, urine, feces	

*The amount of infectious HIV in these fluids is particularly high in individuals during the initial (acute) phase of infection and in individuals with clinical AIDS.

Researchers have developed methods to test for HIV and estimate the amounts of infectious virus present in various body fluids and secretions. HIV can be isolated relatively easily from blood, semen, and vaginal/cervical secretions (including menstrual fluid). When blood and semen are examined closely, the great majority of HIV is associated with infected cells (mostly macrophages) present in these fluids. In blood, if the cells are removed, low levels of HIV are present in the cell-free serum. It has also been isolated from breast milk. With much greater difficulty, the virus has, on occasion, been isolated from saliva, tears, and urine. It has not been isolated from perspiration and feces. The current scientific view is that body fluids and products other than blood, semen, vaginal/cervical secretions, and breast milk contain so little, if any, HIV that they are not of major importance in HIV transmission between individuals.

The relative HIV infectivity of different body fluids and products can be explained in another way, using biological considerations. The fluids and products listed in Table 7-1 differ in the amount of live cells they contain. Blood, semen, vaginal/cervical secretions, and breast milk contain high numbers of live cells. The other body fluids and products (saliva, tears, perspiration, urine, and feces) are completely or nearly completely free of live cells (although they may contain nonhuman cells, such as bacteria). Because live infected cells produce HIV, we would expect fluids with live cells to have higher concentrations of HIV, and this is what many different studies of body fluids have shown.

Stability of HIV

For transmission of HIV infection to occur, infectious virus must survive long enough to pass to a susceptible person and infect target cells. In HIV, the virus particle (see Chapter 4) is actually a very fragile one, as discussed earlier. As a result, the virus quickly becomes inactivated when exposed to the drying effects of air or light. It is also quickly inactivated by contact with soap and water.

As mentioned, much of the infectious HIV is associated with cells (macrophages and T-lymphocytes). In blood or semen, cells maintain infectious HIV as long as they themselves are alive. Thus, intravenous transfusions of HIV-infected blood or sexual intercourse involving HIV-infected individuals efficiently transmits infection because live cells are passed. On the other hand, if blood or semen is allowed to dry, the cells die quickly, and the HIV infectivity is lost.

Targets for HIV Infection

At the cellular level, HIV infection requires the presence of virus receptors on the cell surface. As described in Chapter 4, the receptor for HIV is the CD4 surface protein, which is present only on T_{helper} lymphocytes, macrophages, and the Langerhans cells. Thus, these are the cells that could possibly become infected in an individual exposed to HIV. These cells are most abundant in the blood. Consequently, activities that introduce infectious HIV, either as infected cells or as free virus, into the blood of an uninfected individual have the potential to result in infection. For example, sexual intercourse can result in damage to or microscopic tears in the mucosal linings of the female genital tracts or the male or female rectum. These tears can allow passage of blood or semen into the circulatory system of the uninfected individual. In addition, as described previously, macrophages and Langerhans cells are also present at the mucosal surfaces of the rectum and genital tract, and they potentially can be infected directly without the necessity of virus entry into the bloodstream.

Other potential targets for HIV infection are the oral cavity and the throat. Like the genital tract and the rectum, the throat's mucosal lining contains macrophages and Langerhans cells. In certain sexual activities, such as oral sex, semen can be exchanged orally from one person to another. Consequently, there is the theoretical potential for infection. The epidemiological reality, however, is that oral sex is not a primary mode of transmission of HIV, as we shall see later. The explanation for this may be that there is less physical trauma associated with oral sex or that chemical and physiological features of the oral cavity reduce the efficiency of transmission. This case demonstrates the need to combine theoretical predictions from biology with epidemiological data about incidence rates to understand fully the risks of HIV transmission. It is this topic to which we now turn.

Modes of HIV Transmission

We are now ready to analyze the modes of HIV transmission from person to person and the relative risks associated with different modes. In making our assessment of risk, we rely on both the plausibility of HIV transmission, based on theoretical biological analysis, and the empirical facts associating documented HIV transmission with various modes, drawn from epidemiological studies. Together, these two sources of information permit us to categorize activities and behaviors according to the degree of their association with HIV infection.

Activities Not Associated with HIV Transmission: Casual Contact

Because HIV is so fragile outside the body, transmission requires direct contact of two substances: fluid containing infectious HIV from an infected person and susceptible cells (usually via the bloodstream) of another person. Because of the absence of this type of direct contact, a large group of interpersonal activities and behaviors, generally referred to as casual contact, have no measured association with HIV transmission (see Chapter 6, p. 109) and therefore pose no risk for HIV infection.

What do we mean by *casual contact*? Casual contact includes all types of ordinary, everyday, nonsexual contacts between and among people. Shaking hands, hugging, kissing, sharing eating utensils, sharing towels or napkins, using the same telephone, and using the same toilet seat are a few examples of casual contact. It is impossible to list all types of casual contact here, but you can analyze or make predictions about others, keeping in mind the need for direct contact with body fluids containing infectious HIV. For example, consider the possibilities of waterborne or airborne transmission. Because HIV is quickly inactivated outside the body, it cannot survive in the open air or in water. Consequently, we would predict that there is no risk in sharing the same physical space with a person with AIDS or in swimming in the same pool. Epidemiological evidence supports this conclusion: There is no measured risk of HIV transmission.

Activities Associated with HIV Transmission: Birth, Blood, and Sex

HIV transmission occurs when there is direct contact between HIV-tainted fluid from an infected person and the bloodstream or a mucosal lining of another person. Epidemiological data point to three modes of HIV transmission from person to person: via birth, via blood, or via sex. These are listed in Table 7-2 and discussed in the remainder of this chapter. For most people, the last mode of transmission—via sex—is the most likely, and we discuss it at length. First, however, we briefly discuss the other two modes.

Table 7-2	Modes of HIV Transmission
1. Birth: Perinatal transmission from an infected mother to her gestating infant.	
2. Blood: Transmission from an HIV-infected source to the bloodstream.	
3. Sex: Intimate sexual contact with an HIV-infected person.	

1. Birth: Perinatal Transmission from an Infected Mother to Her Gestating Infant

This mode of HIV transmission brings together a source of HIV (in the bloodstream of an HIV-infected woman) and a potential target (the bloodstream of a developing fetus) in a protected environment (the mother's womb). The mother's and child's bloodstreams are separated by the placenta, which prevents exchange of cells but not of nutrients. During the third trimester of pregnancy, however, small tears sometimes occur in the placenta, which can lead to entry of cells from the mother's bloodstream into the child's. In addition, during birth, the child frequently comes into close contact with or swallows the mother's cervical/vaginal secretions or blood due to the bleeding normally associated with delivery. Although reports vary, a good estimate is that there is about a 23% chance that a child of an infected mother will be infected if there is no antiviral treatment (see Chapter 5, p. 89). Intensive treatment, however, of the mother during pregnancy and delivery and of the infant during the first 6 weeks of life can reduce the transmission rate to 2%. Less intensive treatment, namely starting medications during labor, can even make a big difference, reducing the transmission rate from mother to infant to less than 10%.

2. Blood: Transmission from an HIV-Infected Source to the Bloodstream

This mode of transmission relates to receiving blood or blood products. It can occur in two main ways: by receiving a transfusion of HIV-infected blood or by injection with an HIV-contaminated syringe (either accidentally or incidentally).

▓ Receiving a Transfusion of HIV-Infected Blood

Because a transfusion involves placing foreign blood or blood products directly into the recipient's bloodstream, the necessary conditions for HIV transmission are present: direct contact of potentially infected fluid with susceptible cells in the recipient. Before 1985, when screening of the blood supply for HIV by the antibody test was begun (see Chapter 4, p. 58), the sufficient condition for contracting AIDS was present: HIV-infected blood for transfusion. Even then, however, the risk was low that the blood or blood product involved in a transfusion was infected—except for hemophiliacs who required a clotting factor extracted from the blood of many different donors.

Now the situation is very different. All donations of blood and blood products in the United States are screened in three ways related to HIV: for HIV-1 antibodies, for HIV-2 antibodies, and for p24 antigens. All blood and blood products that test positive for these tests, or for any of about a half dozen other tests that are done (for diseases including hepatitis and syphilis), are discarded. These multiple screenings have greatly reduced the risk of HIV transmission but have not entirely eliminated it. Why is there still some risk? First, the screening tests and the people administering them are not perfect; errors can inadvertently occur. Second, and more likely, there is also a possibility that antibodies have not yet developed in a recently infected donor, so the tests could not detect these.

Blood donation centers have developed methods to reduce the risk even further (Figure 7-1). In addition to routine screening using the tests discussed in Chapter 4, pp. 58–61, centers have developed information campaigns that discourage blood donation from those who might be infected. Procedures also have been established to permit donors, particularly those who may feel pressured during a work-associated blood drive, to indicate confidentially that their blood should not be used. For example, the American Red Cross blood donation centers have used a special card describing a procedure that must be followed by all potential donors. The card lists nine groups of people who should not give blood and then describes a confidential procedure that all donors must follow, involving bar code labels indicating "transfuse" or "do not transfuse." People in one of the nine groups (e.g., drug users, men who have had sex with men since 1977) are to remove the "do not transfuse" bar code tag and place it on

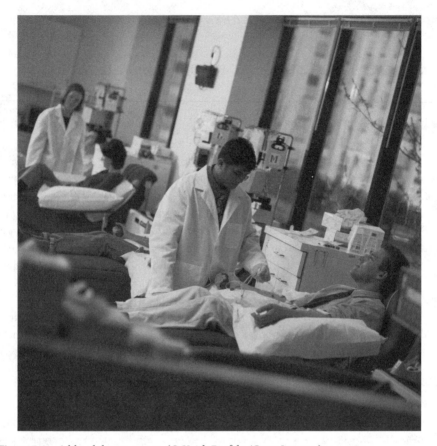

Figure 7-1   A blood donor center. (© Keith Brofsky/Getty Images.)

another card. Those not in one of the listed groups remove the "transfuse" bar code tag and place it on the card. The bar code tags look identical to the casual observer but not to the optical scanner, which later identifies the blood to be rejected. With procedures like these, the risk of receiving infected blood or blood products from a transfusion has become even smaller.

Before we conclude this section, it is important to note three points. First, we have been analyzing the potential risks of receiving a blood transfusion, not of donating blood. There are no risks of HIV transmission from donating blood. The donor's blood is the only potential source of HIV in this situation: If there is no HIV in that blood, there is no other source of the virus. Second, receipt of an organ transplant is a possible source of HIV if the organ donor is HIV infected. Like blood donations, organ donations are tested for HIV, so the risk is quite small. Sperm donations also could be HIV infected and are screened for HIV. Third, the Red Cross's bar code program focuses on groups of individuals and thus risks making the error noted at the beginning of this chapter: focusing on groups that individuals are part of (and thus stereotyping) instead of on their behaviors (which are the important causal factors). Their instructions, however, focus the donor on his or her behaviors that might be associated with these groups, and they leave the final decision to the donor, who understands his or her actual risk-behavior situation best.

■ Injecting Oneself with HIV-Infected Blood

There are two ways in which HIV-contaminated blood in needles can lead to transmission: when needles are shared during injection drug use and through accidental needlesticks between an HIV-infected individual and a health worker. In both cases, someone with HIV-infected blood must be involved for there to be a source of HIV.

In the case of injection drug use, two necessary elements are present: a source of HIV-infected blood and the target of direct injection of that blood into the bloodstream. During the process of injecting the drug, an individual draws blood into the syringe to be sure that the needle is in a vein. Infected blood, therefore, can be mixed with the drug solution. If the syringe is then passed to another individual and inserted into his or her body, contaminated blood from the previous person can be passed into the bloodstream as part of the drug solution.

At first, this mode of transmission may appear contradictory in that HIV is removed from the body and isolated before being passed to another individual. This process occurs, however, in the special context of a protective container—the closed container of the syringe—where blood cells and virus are not exposed to the environment. In addition, it is generally done in a very short time, usually within seconds or, at most, minutes. Consequently, the blood cells remain alive and, with them, the HIV.

Prevention of this mode of transmission involves breaking the link between individuals via the syringe. Injection drug users are encouraged first not to share needles. Some cities provide free sterile needles so that limited syringe availability is not an issue, thereby reducing the chances for harm. Alternatively, injection drug users are encouraged to clean their needles between administrations, using a bleach solution.

The other mode of HIV infection is via accidental needlesticks by health workers. On occasion, health workers, in emergencies or in the process of medical laboratory work with HIV-infected people, have accidentally stuck themselves with potentially contaminated needles. There are a total of about 800,000 needlesticks in the United States every year from syringes, various intravenous needle assemblies, and blood-draw equipment; of these, about 16,000, or 2%, are with HIV-contaminated devices.

The risk of contracting HIV from a contaminated needlestick is about 1 in 200 (one-half of one percent, or 0.5%). Consequently, the risk that a healthcare worker will accidentally stick himself or herself with an HIV-contaminated needle and then develop HIV is 1 in 10,000 (one one-hundredth of one percent, or 0.01%). Administering antiretroviral drugs soon after exposure (between 2 and 36 hours) and then continuing the regimen for 4 weeks can reduce this small risk even further. Based on a Centers for Disease Control and Prevention study in 1995, the reduction in risk of infection may be as much as 81%.

To put these figures in a larger perspective, it is instructive to compare the transmission rates of HIV and the hepatitis B virus (a virus transmitted by similar routes as HIV) by needlestick injuries. The rate for transmission of HIV via a contaminated needlestick is approximately 0.5%. In contrast, the rate of transmission for hepatitis B virus via a contaminated needlestick is between 6% and 30%.

The greatest risk of HIV infection for healthcare workers is by an accidental needlestick, but, as the data presented here show, this risk is, in fact, quite low. Nonetheless, the risk does exist, as does the relatively greater risk of contracting other infectious diseases, such as hepatitis, via needlesticks. Consequently, during clinical procedures, health workers (who should also be wearing gloves) have been advised to discard used needles directly rather than recapping before discarding. In addition, new needles have been designed that make accidental sticks more difficult.

3. Sex: Intimate Sexual Contact with an HIV-Infected Person

For most people in the general public, this mode of transmission is the most likely source of HIV infection. The risk differs depending on the particular sexual practice, the frequency of the practice, and the HIV status of a sexual partner. We cannot, therefore, categorize particular sexual practices with certainty in terms of their HIV risk. The degree of risk of any particular sexual behavior differs from person to person. Individual risk assessment, however, is usually a difficult task, based on incomplete and sometimes unknowable data (e.g., the HIV status of a new sexual partner).

To make this task somewhat easier, we can discuss what we know from the theoretical and epidemiological perspectives. Together, the data from these perspectives complement each other and provide useful information for judging the relative risk of various sexual practices.

From the theoretical perspective, we know that we need two critical elements together: HIV-contaminated body fluid (in particular, blood, semen, or cervical/vaginal secretions) and direct contact of this fluid with a target site. The riskiest sexual practices, therefore, are those in which HIV-infected blood, semen, or cervical/vaginal secretions from an infected person come in immediate and direct contact with the bloodstream or mucous membranes of another person. These practices include vaginal intercourse between a man and a woman, anal intercourse between a man and a woman, and anal intercourse between two men. In all these practices, semen from the man is deposited into the vagina or rectum—both sites of HIV-target cells in macrophages and Langerhans cells and also sites where small tears frequently occur during intercourse, exposing the bloodstream.

At the other end of the spectrum, the least risky sexual practices are those where HIV-infected blood, semen, or cervical/vaginal secretions do not usually come into contact with target sites. These practices include masturbation by a male onto the unbroken skin of a partner and dry kissing (closed-mouth kissing). In the case of male masturbation, although a potential source of HIV is present in semen, a target site is not because unbroken skin does not present an HIV target. In the case of dry kissing, there is usually no source of HIV, since neither blood nor semen is generally present in the mouth, and the saliva of HIV-infected people has been shown to contain little or no HIV. With no source for HIV, there would be no risk.

On this spectrum of risk, we can anchor the ends of potentially risky sexual behaviors but cannot precisely anchor other groups of sexual practices or consider every possible case that could arise. For example, what if two people are dry kissing and one has a cut on the lip: Is there a risk of HIV infection? Or, what if a man masturbates onto chapped skin: Is there a risk of HIV infection? The answer to both questions is "possibly." Here is where the epidemiological evidence is useful.

We know that across groups of people, those who frequently engage in particular sexual practices are more likely to become HIV infected. The sexual practices listed here at the "riskiest" end of the spectrum of risk (vaginal intercourse with an HIV-infected person without a condom, anal intercourse with an HIV-infected person without a condom) have been shown, through epidemiological data, to be highly associated with HIV infection. The two practices from the "least risky" end of the spectrum (dry kissing and masturbation by an HIV-infected male onto the unbroken skin of a partner) have not been shown, through epidemiological data, to be associated with HIV infection.

Epidemiological data also provide clues to the relative infectivity of other sexual practices. Wet kissing (open-mouth kissing with exchange of saliva) has not been shown to be associated with HIV transmission. This makes sense from a biological perspective because we know that saliva of an HIV-infected person contains little, if any, HIV, so there is not a likely source of infection. Oral sex performed on an HIV-infected man or woman by either a woman or a man has not been strongly associated with HIV transmission, although there are some reported cases of transmission via this sexual practice. From a biological perspective, we can see why transmission might be possible, if HIV-infected semen is deposited in the mouth and throat or possibly into the bloodstream via small tears in the mouth. Nonetheless, there must be other chemical or physiological factors (e.g., the acidity of the mouth) that provide some barrier to HIV transmission because the epidemiological data do not show oral sex to be highly associated with HIV transmission.

From our analysis of the relative risk of various sexual practices based on biological and epidemiological considerations, we can both place the sexual practices on the spectrum of HIV risk and also see ways to reduce the risks of all sexual practices that could involve some risk. Abstinence from sexual relations clearly reduces the risk of transmission to zero: There is no source and no target. Because most people do not choose abstinence, another option is to have sexual relations of the least risky types. If they choose riskier sexual practices, they can reduce the risks by placing barriers between potential sources of HIV infection and potential targets. For example, they can use condoms during vaginal and anal intercourse to reduce the risk of HIV infection by containing potentially infected semen within the condom and preventing its contact with target sites in the vagina or rectum. Condom use during oral sex on a man also provides a barrier between potentially infectious semen and the target sites in the mouth and throat. During oral sex on a woman, a dental dam (a 3- to 4-inch square piece of latex) placed over the vagina also provides a barrier for source-to-target site contact.

These protective methods are not 100% effective. Condoms can have holes and can leak, although this is not at all frequent. Studies are regularly done on condom reliability, and condom manufacturers pay close attention to quality-control procedures. Putting on or using a condom in the wrong way can be another cause of ineffectiveness (for example, not rolling the condom completely to base of the penis). In fact, most studies indicate that condom misuse is a bigger cause of problems than are technical manufacturing issues. Consequently, improved knowledge and practice by condom users will result in very effective protection, not only of HIV transmission but also of pregnancy.

Properly used, therefore, condoms provide very good protection for most people. A condom should be fresh and made of latex (not of natural products). The condom must be placed on the man's erect penis before any penetration because preejaculatory fluid has been shown to contain HIV in an HIV-infected individual. Space should be left in the tip of the condom for the semen that will soon be ejaculated, and the condom should be unrolled completely to the base of the erect penis. If a lubricant is used during intercourse, it should be water based; grease- and oil-based products destroy latex.

Condoms or lubricants with nonoxynol-9, a spermicide, used to be recommended; recent studies, however, indicate that nonoxynol-9 does not kill the virus and may even cause irritation that could make HIV entry more likely. Consequently, condoms without nonoxynol-9 but with a water-based lubricant are now recommended. The condom must stay in place at the base of the penis until the penis is withdrawn from the vagina or rectum; this is best done before the man's erection fades and the penis is flaccid and separated from the stretched condom.

Each person must analyze his or her own sexual practices and take the precautions necessary for protection from HIV. The guidelines described in this chapter for self-protection are similar to those advocated by the U.S. Surgeon General, the U.S. Centers for Disease Control and Prevention, and many local AIDS prevention programs. The final decisions about individual risk assessment and management are made differently by each of us. Although we will never have all the data we need to make perfect decisions, we can use information from biological and epidemiological studies to assess the risks of various sexual practices. These assessments are best made before sexual activity, when our thinking is less affected by volatile emotions and judgment-confusing substances (alcohol or drugs), which are sometimes associated with sexual behavior for some people. Because sexual activity usually involves two people, it is necessary to think about HIV risk assessment and make decisions on HIV risk management together with your sexual partner before sexual activity. Then, those who choose to have sexual relations will be ready to enjoy the sexual experience more, knowing that they have taken the necessary precautions to lower the risk of HIV transmission.

---

http://biology.jbpub.com/fan/aids/6e/

Connect to this book's website: http://biology.jbpub.com/fan/aids/6e/. The site features summaries of the main points from each chapter, links to important AIDS-related websites, and short-answer-style review questions for each chapter.

CHAPTER 8

Individual
Assessments of
HIV Risk

Introduction to Individual Decision Making and Action

Risk Assessment
Normative Model
Subjective Probability Model

HIV Testing and Risk Assessment
Nature and Accuracy of the HIV Test
Testing Context: HIV Counseling as Part of the HIV Test
Testing Options

We begin now to look at HIV/AIDS from the individual's perspective. (For convenience, the inclusive term *HIV/AIDS* is used whenever the discussion may apply both to HIV and to the disease state resulting from HIV infection [AIDS].) In this chapter, as well as in Chapters 9 and 10, we address the following questions: How do individuals assess their own risk of HIV infection? What contribution can the HIV antibody test make to this assessment? What factors affect individuals' knowledge, attitudes, intentions, and behaviors related to HIV/AIDS? How can these factors be incorporated into HIV/AIDS prevention programs? How do individuals—both those with HIV/AIDS and those who are not infected—live with the changing realities of AIDS? In this chapter, we focus on the first two questions related to assessing HIV risk. Following an introduction to the general process of individual decision making, we consider two main topics: the general issue of individual risk assessment and the specific issue of HIV testing.

Introduction to Individual Decision Making and Action

Each of us continually makes decisions that affect our actions. Many of these decisions are minor and relatively unimportant ("What will I have for breakfast?"), but others

135

are major and have significant consequences ("What job will I take?"). Still other decisions may seem minor at the time ("Do I wear a condom if I have sex?"), but they can have important outcomes (for instance, decreasing or increasing the risk of HIV transmission).

A general model underlies all these cases, with four basic steps (ordered temporally):

1. Knowledge
2. Attitude
3. Intention
4. Behavior

Knowledge involves information collection and synthesis. *Attitude* condenses this information into a conclusion. *Intention* involves a readiness to take action, and *behavior* is the action actually taken.

This simplified model separates the parts of the decision-making and action processes so that we can understand the factors that affect each step related to HIV/AIDS. This chapter focuses primarily on the first two steps for individuals assessing their own HIV risk; the next chapter expands the picture to the other two steps. As we will see, different factors affect the four steps. For example, on a particular topic or issue, an individual's openness to information in the "knowledge" step is affected by his or her other attitudes and past and present actions on related issues.

In the case of HIV/AIDS, the decision-making process is a particularly complex one because of the probabilistic nature of HIV risk information and HIV risk. There are some matters in our lives that are clear-cut. For example, if we do not eat and drink, the outcome is clear: We will die. Many other matters, however, are not so clear. If we spend a lot of time outdoors in the sun, we may—or we may not—contract skin cancer. It is the "may or may not" aspect that makes such decisions probabilistic—less than 100% but more than 0% certain. HIV is one of these probabilistic issues. In the following section, we discuss several factors that affect individuals' assessment of information, such as that related to HIV transmission.

Risk Assessment

Consciously or unconsciously, each of us makes *risk assessments* throughout the day as part of our decision making. Should I choose the salad or the fries with my sandwich, knowing that the salad is better for me? If I select the salad, do I choose the tasty but fatty dressing (knowing that extra calories and fat are not good for my heart and arteries), or do I decide on a little lemon juice instead (knowing that it has almost no calories and fat but also less taste)? Likewise, confronted with a need to take action related to our risk of contracting HIV, we have to weigh information about risks before we make decisions. Karen and I are going to have sex: Could she have the HIV virus? Do I have the virus? What are the HIV transmission potentials for different sexual acts? How do different prevention measures (such as using a condom) affect the HIV transmission potentials? Much of this risk information is probabilistic and therefore is usually considered in special ways. What is probabilistic information and what are the "special ways" used to consider it?

Probabilistic information contains an estimate related to an issue or topic. The estimate may be about likelihood, frequency, or prevalence. For example, the weather bureau's announcement that there is a "40% chance of showers" is probabilistic information. Another example is the statement from the United Nations' 2008 Report on the Global AIDS Epidemic that 45% of all new HIV infections in adults is among those aged 15 to 24 years, highly disproportional to this age group's 20% share of the worldwide population. These statements provide some information, but they do not tell us exactly what we would like to know in each case to make a personal decision and take action. For instance, in the weather example, if we know that there is a 50% chance of showers, this information does not directly lead to a clear personal decision about taking an umbrella or canceling our tennis game. In this example, if the probability were more extreme (for example, a 5% chance of rain or a 95% chance), the situation would be clearer, and it would be easier for us to use this probabilistic information to reach conclusions and take personal actions. In the HIV example, if we are in the age range of 15 to 24 years, we cannot simply generalize from the 45% infection figure to our own group of similarly aged friends, acquaintances, and potential sexual partners.

In the case of the probabilistic information about the high number of new HIV infections among young people, a person in this group faces additional challenges in trying to interpret its meaning and personal importance in terms of his possible behaviors. To highlight these challenges, consider the risk assessment facing a particular member of this group, a college student named Marc who is considering sexual contact with another member of the group, Amy, who is also a college student. Marc's question is whether Amy is likely to be HIV positive. The statistic that "about 1 in 2 new HIV infections are among young people" gives Marc a limited amount of information but does not tell him whether Amy is HIV positive. Even if he had more specific national information about the particular group of young people of which he and Amy are members, including their HIV-positive rate, Marc will need to consider other information, and it is also probabilistic. This information includes the degree of risk for HIV transmission associated with particular sexual practices and the degree of protection against HIV transmission afforded by various products (for example, condoms). Some of this latter probabilistic information is not even available in numerical form. For example, in Chapter 7 we talked about less risky or more risky sexual practices; we cannot assign percentages or proportions to particular practices or even to groups of practices beyond these general characterizations.

How is Marc (our hypothetical risk assessor) to make conclusions from all this probabilistic information about what will happen between him and Amy? This question moves us to the issue of how we make decisions. Most of us believe that people generally make decisions rationally (using a normative model, which is discussed next). In fact, people are often "predictably irrational" in their decision making, to use a term from behavioral economics.[1] This is not to say that people are intentionally or capriciously irrational; actual decision making, it turns out, is subjective and involves various shortcuts (*heuristics*) to apply probabilistic, general facts in concrete, specific circumstances.

To understand the difference between how most of us think we and others make our decisions and how we actually do, let's begin with the "facts." Probabilistic facts are considered in two main ways: according to a normative judgment model or according to a subjective probability model.

Normative Model

In the *normative model*, probability information is weighted according to statistical rules to reach conclusions. A detailed understanding of these statistical rules is not necessary for our purpose. What is important, however, is a general understanding of how the normative model is applied by scientists, which can be conveyed by an example.[2] In this illustration of the normative model, the decision to be made is this: What is the occupation of the person described in the following sentences? "Steve is very shy and withdrawn, invariably helpful, but with little interest in people or in the world of reality. A meek and tidy soul, he has a need for order and structure and a passion for detail." The list of possible occupations includes farmer, salesperson, airline pilot, librarian, and physician. What is Steve's occupation? In making his or her decision, a scientist using the normative model would consider the base-rate frequency of each occupation in the population. If one occupation is much more common in the general population than the others, the likelihood is that Steve is from that population, and the best choice, therefore, would be that occupation. At the time of the study, there were many more farmers than salespeople, pilots, librarians, or physicians, so farmer would have been the choice using the normative model of base-rate frequency. If we update the base-rate frequencies of these occupations in the United States today, the most common occupation among these five categories is salesperson (with a frequency that is more than twice as great as all the other four occupations combined); consequently, salesperson would be the best choice now of Steve's occupation using the normative model rule.[3]

Other normative model rules can come into play when scientists make decisions. One such rule is the independence of random or chance events. If we toss a coin four times, each toss is independent. The outcome of the second toss (heads or tails) is independent of the first toss and the subsequent tosses. If the first three tosses of the coin result in heads, heads, and heads, the scientist using this normative model rule will predict an equal chance of either heads or tails for the fourth toss because each toss is independent. (Put another way, the coin cannot "know" the outcomes of the previous tosses.)

While reading each of these examples, you may have made your own decisions about Steve's occupation or about the outcome of the fourth coin toss, and these decisions likely were not what the normative model would predict. Many people will guess that Steve is a librarian—even with the additional information about the base-rate frequencies—and that the fourth coin toss will more likely be tails than heads. These conclusions are common, and they cannot be labeled "irrational" because there is a set of rules being used. The set of rules, however, does not come from the normative model but instead comes from a different model: the subjective probability model.

Subjective Probability Model
Most of us do not and cannot use the normative model for most of our decision making. In most cases, even if we wanted to use statistically based normative rules, we do not have all the necessary data and, in some cases, cannot have all the data because some probability data keep changing. (As an example, think about the variations in the weather reports we receive over the day; what was a 40% chance of rain in the morning may be a 60% chance by mid-afternoon.) Given the question above about Steve's occupation, most of us do not know how common different occupations are in the general population, in our geographical area, or elsewhere. So, we use other information that we have at hand—or rather, in our heads—and use what are known as "judgment heuristics."

Judgment Heuristics
The rules most of us use to make decisions based on probabilistic information are known as *judgment heuristics*. These are rules of thumb, identified in a variety of research studies, that provide shortcuts in processing probabilistic information. For most of us, we use these rules unconsciously and almost automatically, unaware that we are doing so. To see how this happens, we will consider the three most commonly used heuristics in more detail: representativeness, availability, and anchoring.

■ Representativeness
One general decision-making rule of thumb that most of us typically use to assess probabilistic information is representativeness. If two objects or items are highly similar, based on what we know—or believe we know—about them, we then conclude that one item or object is representative of the other. Based on this conclusion, we make decisions.

Think back to the example of matching the description of Steve (". . . very shy and withdrawn . . . little interest in people . . . a meek and tidy soul. . . .") and his likely occupation (farmer, salesperson, airline pilot, librarian, or physician). Using representativeness, most people compare the details of the description given to them with the stereotypes they have for each of the occupations, note a high similarity with the occupation of librarian, and conclude that Steve is a librarian.

In many cases, the representativeness heuristic provides good—or good enough—decisions. Guessing Steve's occupation correctly or incorrectly is not an issue of major importance to us, so if we make an error, the consequences are not great. For other issues, however, particularly those personally related to us with significant consequences, the right outcomes are much more important. In these cases, the representative heuristic can lead to errors in judgment because it ignores factors and facts that are relevant to judgments of probability.

Consider an example involving Lan, a college student who is estimating the risk of contracting HIV from her fellow students. In making her judgment, she compares the similarity of (that is, the representativeness between) her peer group and two groups she understands to have high HIV infection rates: male homosexuals and intravenous drug users. She sees little similarity. In addition, she compares her fellow students with the general population, a group that has a very low HIV infection rate, and sees many similarities. Based on assumed representativeness, Lan concludes that her fellow students, like the general population, have a low HIV infection rate. Taking this one step further, she decides that her risk of HIV is very low.

We could argue, however, that she should use a different group that is more comparable to her: young people. We know that a disproportionate number of new HIV infections are now among young people, so the HIV risk is not low. Lan's conclusion that her risk is very low underestimates the risk of HIV infection from her fellow students. It is important to understand that Lan is not intentionally distorting the data and making irrational decisions. Instead, she is making comparisons that seem appropriate to her, drawing upon her understanding of the data, and then using the representativeness heuristic to generalize about probabilities of HIV infection rates. Note that Lan is using her own understanding of the data in making a decision—her subjective assessment of the probabilities. (In some cases, additional factors are present—for instance, a strong emotional factor—which may distort the subjective probability assessment. The influence of some of these other factors is considered later in this chapter and in Chapter 9.)

Consider another example related to HIV/AIDS in which decisions made on the basis of the representativeness heuristic could lead to problems of a different sort, this time in the overestimate of HIV transmission risk. Miguel works as a clerk at a health clinic that has some HIV-positive clients. He is considering his degree of HIV risk from his contact with these clients. He believes that viruses, like the common cold, can be spread by casual contact (e.g., kissing someone with a cold, breathing the same air after someone with a cold has sneezed repeatedly, touching the same surfaces). Miguel also knows that HIV is a virus. Using the representativeness heuristic, he groups HIV with other viruses and concludes that, like other viruses, the risk of HIV through casual contact with HIV-positive clients is relatively high. Based on this, he restricts his behavior with HIV-positive clients in the clinic.

As in the previous example, Miguel is not making "irrational" decisions. Instead, he is using information he believes to be true to help him make decisions about the probabilities of events that, in his mind, are related. From a scientific standpoint, however, his inclusion of HIV with other viruses is incorrect in terms of similar modes of transmission, and therefore his decision about the risks associated with casual contact is incorrect; the actual risks do not justify the restrictions he made in his behaviors with HIV-positive clients at the clinic. Nonetheless, using the representativeness rule of thumb and what he believes is correct, he groups together similar objects and then makes a decision about risks of HIV transmission. In this case, we could provide useful input for Miguel's decision by giving him correct information about how HIV is transmitted.

We want to highlight two aspects of the representativeness heuristic, both illustrated in the two examples. First, the application of the representativeness rule of thumb, like other rules of thumb discussed next, may occur almost unconsciously. In all likelihood, if our hypothetical examples came to life, neither Lan nor Miguel would consciously outline the decision task in the way we described it. Instead, they would move through the steps almost instantaneously. Think about how you likely reacted to the initial description of Steve and the list of occupations. Your application of the representativeness heuristic to label him a librarian probably occurred almost instantly. It is doubtful that you carefully considered the details of his description and then self-consciously compared them with your own description of characteristics of farmers, salespeople, airline pilots, librarians, and physicians. This is why we call these thought processes heuristics or rules of thumb: They are quick shortcuts to probability decision-making tasks.

Second, it is the *perception* of reality—not reality itself—that determines how someone will decide. Thus, this approach to decision making is called the *subjective* probability model. If a person is very well informed and knows the data, his or her perception may match reality, and the decisions may be more objectively based. Otherwise, it is his or her understanding of reality, rather than the reality itself, that becomes the source for decision making, and the decisions will be more subjectively based. In our example, Miguel perceived HIV to be like other viruses in the way they are spread. We know that HIV is not spread through casual contact; that is the reality, based on extensive epidemiological data and theoretical reasoning. Miguel's perception of reality, however, is that HIV is spread through casual contact. Consequently, he makes decisions that are logical, given this "reality." If we assume he is illogical or stupid, we miss the important point that he, from his viewpoint, is deciding rationally and intelligently. We also miss an important opportunity to understand how we could change his knowledge—part of his reality—and thus change the conclusions he reaches.

■ Availability

Another important rule in our decision making about probabilities is the availability in memory of items or objects. We judge an item or object that is present in our memory to be more probable than one that is not present or only weakly so. The two main contributors to the availability of an item or object in our memory are familiarity and salience.[4] *Familiarity* is the frequency of occurrence of an item or object in our memory, and *salience* is the distinctiveness or vividness of an item or object in our memory, apart from its frequency.

Several examples will help to distinguish familiarity and salience. In the first example, the decision-making task is to estimate the percentage of the U.S. population who own cars. This task is given to Paulette, who works and lives in a large U.S. city, and to Tran, who lives and works in a rural area with only a few small towns. Cars are much more familiar in memory to Paulette, who works, lives, and moves around them all day, than they are to Tran, who sees few cars during a day. Because of these differences in familiarity, Paulette will probably overestimate the percentage of the U.S. population with cars and Tran will underestimate it.

Salience is distinct from familiarity. Even if an event or object does not occur frequently, the presence of the event in a person's mind is increased if it is dramatic when it does occur. This makes the event more available in memory and thereby increases a person's perception of its probability.

The next example relates to airplane crashes, which are very dramatic events made even more so by vivid media images. Most people overestimate the probability of an airplane crash by a wide margin. A primary reason for this is the saliency that an airplane crash has for most people on the rare occasion when it occurs. News pictures of flames leaping from an airplane's wings and windows and of passengers fleeing through smoke and fire or of debris floating on the ocean's surface after a plane crashes into the water create searing images in most people's minds. The saliency of these images makes them available in memory for a long time, increasing personal subjective estimates of the probability of an air crash occurrence. Contrast this situation with the image most people have of a bicycle crash, if they have an image at all. We would expect people to judge their risk of having a bicycle crash as very remote, given the very low salience of this event and its lack of availability in memory.

The relevance of this factor of availability is easy to see for the case of HIV risk assessments. Those who frequently see or have contact with people with AIDS give higher estimates of the general prevalence of HIV/AIDS compared with those who have no experience with AIDS/HIV. For example, surveys of citizens in San Francisco, one of the hardest-hit U.S. cities in terms of AIDS as well as one of the most active cities in terms of AIDS prevention and treatment, show that they overestimate the prevalence of HIV/AIDS. Images and information about HIV/AIDS are more familiar to them and therefore more available in memory when they are making estimates of HIV infection or AIDS.

An example of the effect of salience is the increased attention to AIDS that occurs when a well-known figure contracts the disease and the related increase in people's assessments of HIV/AIDS risk. The movie star Rock Hudson is credited with bringing the reality of AIDS inadvertently to many Americans for the first time in the mid-1980s. When the news media gave front page attention to his diagnosis of AIDS and the quick progression of the disease, the salience of AIDS increased sharply for many people. A similar effect occurred when basketball star Magic Johnson revealed that he was HIV positive and was retiring from professional basketball. The news media again gave major coverage to the topic, which increased the saliency of AIDS for many young people, especially African American youth. In both the Hudson and Johnson cases, healthcare workers reported sharp increases in contacts from people who had reassessed and increased their judgment of personal risk for HIV/AIDS. Indeed, in talking about the pattern and frequency of HIV tests, some healthcare workers refer to the "Magic Johnson spike"—the greatly increased number of people who sought HIV testing right after Johnson's dramatic announcement.

From our understanding of the factors that affect risk assessments, we can explain these changes in personal risk assessments: Major media coverage results in an increase in saliency about AIDS, which in turn causes some people to increase their assessment of the presence and risk of HIV/AIDS and of their personal perceived risk of HIV.

▓ Anchoring

The final factor we will consider in explaining the reasons for our subjective probability assessments is anchoring. *Anchoring* is the way in which the starting point for our assessment—that is, our initial estimate or base—affects how we adjust subsequent estimates.

The following example illustrates this factor well.[5] Subjects in a study were asked to estimate the percentage of African countries in the United Nations. They received an arbitrary number between 1 and 100 (by spinning a wheel of fortune) and then judged (1) whether the percentage of African countries was higher or lower than this initial number and (2) what the actual percentage was, moving up or down from the initial number. Different groups were, in fact, given different starting points for their estimates, and these seemingly arbitrary initial numbers significantly affected judgments. Groups that received 10 as the initial number gave average estimates of 25% for the number of African nations in the United Nations. In sharp contrast, groups that received 65 as the initial number gave average estimates of 45% for the number of African nations in the United Nations. These variations occurred even though the subjects understood that the initial numbers were simply random starting points, and the variations persisted even when subjects received payments for accuracy. The initial anchoring set a frame that confined the subjective probability judgment.

For the case of HIV/AIDS risk assessments, the effect of anchoring is demonstrated in the errors people make in assessing the overall risk attributable to a series of events, each with a different probability. If the first event has a high probability attached to it, it increases the assessment of the probability of the other events. As an example, consider this series of sequential events: Brad meets Monica, whom he knows is an HIV-infected person; Monica cuts herself on her arm and starts to bleed; Brad has an open, fresh cut on his finger; Brad touches Monica's blood with his cut finger. For HIV to be transferred from Monica to Brad, at least all of these four events would need to occur (and a few other events as well, but for the sake of simplicity, we will limit our example to these four).[6] Brad's personal risk assessment challenge is to estimate the likelihood of this series of events causing him to become HIV infected. If Brad is like most of us, he will overestimate the probability, compared with its actual statistical likelihood. The extent of overestimation can generally be explained by the anchoring heuristic. Brad starts his estimation at a high probability (Monica is HIV positive) and then adds to his probability estimate for each of the other three events occurring, even if he thinks one or two of them (for example, getting a fresh, open cut on his finger) is very unlikely. In his mind, he ends with a very high subjective probability assessment of his HIV risk from these four events. In fact, the overall probability of all four events coming together and occurring will be smaller than any one of the individual events by itself (statisticians can demonstrate this easily). This is a conjunctive series of events: A and B and C and D. The reality is that the more events added to a series (if each new event is less than 100% likely), the lower the probability becomes of the overall conjunctive event occurring. The subjective perception for most people, however, is that with more events, probability increases.

Another main type of risk assessment event series is relevant in considering HIV risk: *a disjunctive series*. A disjunctive series is one in which A or B or C or D occurs, in contrast to the *conjunctive series* where A and B and C and D must occur. As an example, consider Christina, who had unprotected intercourse with four different men. Her risk assessment challenge is to estimate the HIV risk from all four of these encounters: How likely is it that she contracted HIV from any one of the men? If Christina is like most of us, she will underestimate the actual risk. We will assume that Christina, knowing she was going to have sex with each man, decided that each was HIV negative. So, Christina started her subjective probability assessments at the very low end with her first partner. Then, for each of the other three partners, whom she also assumes are not HIV positive, she adds a very small additional risk to her assessment of the risk in the overall situation. She ends her subjective probability risk assessment by concluding that there was very little risk. In fact, the probability of a disjunctive series of events (A or B or C, etc.) is greater than the probability of any one event. The more events added to the series, the larger the probability becomes of any one of the events occurring, even if the probability of each individual event is quite small.

From an HIV risk standpoint, then, the more events occurring with some HIV risk potential, the greater is the likelihood of transmission. Stated this way, it seems obvious. But, reread Christina's situation and put yourself in her place, trying to assess probabilities subjectively in a sexually—and emotionally—charged context. The initial event, with a very low risk anchoring, caused the subjective perception of the risk associated with the whole set of events to remain low—probably too low, based on the actual probabilities. It is personally reassuring for her to view the situation this way, but if this misperception fosters continued unprotected sexual contacts, the situation could change for the worse. By understanding why Christina makes her assessments and how these might affect her future behavior, we are in a better position to suggest ways to change Christina's subjective assessment so that she sees the situation and risks more accurately.

These three judgment heuristics—representativeness, availability, and anchoring—help to explain how people handle the difficult task of making subjective probability assessments, of which an HIV risk assessment is a good example. Unfortunately, these judgment shortcuts inadvertently contribute to erroneous risk assessments that are sometimes overestimated and other times underestimated, depending on the judgment task and the heuristic used.

Optimistic Bias

The judgment heuristics we have been discussing are used when we make subjective probability assessments about risk to others and to ourselves. When we make these assessments about ourselves, however, another important factor comes into play: optimistic bias. This is also known as *personal invulnerability*.

Many of us, especially the young, tend to view ourselves as less vulnerable to experiencing bad outcomes. We have an optimistic bias that bad things will not happen to us and, conversely, that good things will happen to us, compared with other people. This optimistic bias extends into many aspects of our lives, from our wager of several dollars on a lottery ticket to our quick puffs on a cigarette. In the former case, some of us believe we are luckier than others and therefore that our lottery ticket will be the winning one; in the latter case, some of us believe our bodies are stronger or healthier than other people's, so smoking "just a little" will not do any damage to our lungs and heart. The scientific evidence, however, does not support these personal invulnerability beliefs: The objective chance of winning the lottery is less than the chance of being struck by lightning, and the harmful effects of smoking are cumulative (indeed, nonsmokers, who do not even smoke "just a little," suffer damage to their bodies from inhaling smoke from others).

Why do we have this *optimistic bias*? It arises from our upbringing and is fostered as we grow. Most of us are raised by our families and caregivers to believe we are special. We are encouraged in this belief as we begin, however tentatively, to deal with the outside world and we are praised and rewarded for our successes. Our early failures are overlooked and blamed on external circumstances beyond our control.

All this leads us to conclude, usually unconsciously, that good things are associated with us and that bad things are not. The world is just: Good things, we believe, happen to good people and bad things happen to bad people, and we are among the "good people." This bias toward optimism is beneficial as we grow because it provides both a reason to continue to attempt activities and a ready explanation for success (namely, that we have the talent or skill) or for failure (namely, that the external circumstances conspired against us this time). For many of us, this bias reaches its peak in our adolescent years when we see ourselves as invulnerable to many threats and risks. At its extreme, the optimistic bias can blind us to realities, cause us to make incorrect judgments and to take foolhardy chances, and, on occasion, even threaten our lives.

In the case of assessing our risk for HIV, the optimistic bias tends to make us underestimate our objective risk for several reasons. First, because the outcomes generally associated with HIV/AIDS are bad, we tend to assume that we are at less risk for bad outcomes than other people are (research has shown this to be the case for people's judgments about a variety of bad outcomes). Second, we assume that the people we are with also are more likely to avoid bad outcomes. These, then, are the "facts" that are in our mind: AIDS, like other bad things, will not happen to me, and it will not happen to my friends, lovers, or spouse. Bad things happen to bad people; bad things do not happen to good people like me and my boyfriend or girlfriend. This perception of the "facts" can lead us to conclude that we are not at risk for HIV from sexual contact with our friends, lovers, and spouse.

The objective facts, however, present a different picture and do not fit neatly into a good person–bad person categorization. Our friends, lovers, and spouses have numerous contacts—some of them sexual—with other people, and we cannot know about all these sexual contacts or about the HIV status of all these other people. Even if we could know all these other people, we would also need to know the HIV status of all the other sexual partners of all these other people. (A further confusing objective fact is the window period associated with the HIV test, which we will discuss. For the moment, however, we have more than enough confusing objective facts.) In short, the task of knowing the objective HIV/AIDS risk associated with friends, lovers, and spouses is impossible. Our best choice, therefore, is to assume that the risk of HIV is present and to take precautions. The optimistic bias, however, pushes us in the opposite direction: AIDS will not happen to me, especially not from sexual contact with the good people who are my friends and who will not have HIV.

In summary, we see that certain psychological factors have a major effect on our understanding and interpretation of the "facts" of HIV/AIDS risk. We use judgment heuristics (representativeness, availability, and anchoring) as shortcuts to assess HIV risk for others and ourselves. These rules of thumb are likely to make our subjective probability assessments different from those made by scientists using the objective data. Optimistic bias plays an important role in our personal HIV risk assessments, causing most of us to underestimate our vulnerability to HIV. Although other factors can affect risk assessments,[7] judgment heuristics and optimistic bias are the important ones for most of us, particularly for our initial assessments.

The most important point about all these factors is that in our assessment of our risk for HIV, each of us views the objective "facts" differently. Our risk assessments are subjective—personal evaluations that differ for each of us. Each of us confronts a different set of realities in assessing our HIV/AIDS risks, depending on our sexual and drug use behaviors. When we receive information about HIV/AIDS, we interpret it in the context of our own particular realities. As we interpret the information, we need to be aware of the risk assessment factors that operate to distort our understanding. We also should be aware of these factors in understanding why different individuals come to different assessments of their risk of HIV.

It should now be clear why risk assessment is an uncertainty task. It is easy to understand, therefore, why most of us would like a quick and definite answer to the question, "Am I at risk for HIV?" At first thought, the HIV test might seem to be just the method to get this answer. As we discuss next, an HIV test can provide useful information, particularly about the effects of past risks, but it does not provide a guarantee of no HIV risk for the present or the future.

HIV Testing and Risk Assessment

Will the HIV test provide the definitive answer to HIV risk? Thuy, for example, might believe that both he and his girlfriend, Maricres, should be tested. If they both test HIV negative, he believes, they can conclude that they are not at risk for HIV and can act on this knowledge. Can the HIV test tell Thuy and Maricres this? To answer this question, we need to review briefly the details of the HIV test.

Nature and Accuracy of the HIV Test
The biological details of the HIV test are described in Chapter 4. Here, we focus on two factors related to the test, which have implications for the information individuals can

obtain from the test and the utility of this information in personal risk assessments: the nature of the test and the accuracy of the test.

The HIV tests that are most commonly and widely used are antibody tests. As described in Chapter 4, this means that the tests do not directly measure HIV. Instead, they measure whether antibodies to HIV have been produced. As with any virus, there is a time period between infection and the production of antibodies. This is called the window period, and for HIV it can be as long as 6 months. At the beginning of the period, an individual tests HIV negative (seronegative), and at the end, when antibodies are being produced, the individual is said to have seroconverted and now tests HIV positive (seropositive).

These realities of the test mean that a negative HIV test does not necessarily prove that someone is free of HIV; it proves only that the individual did not have antibodies to HIV at the time of the test. Because of the window period, someone can, in fact, have the HIV virus but test HIV negative because no antibodies have been produced yet. This is an important consideration in interpreting test results and drawing conclusions about present and future HIV risk.

Two other very remote explanations for HIV test results are presented here for the sake of completeness: An individual may test negative and never produce antibodies yet still be infected (due to a quirk in his or her immune system), or an individual may test positive for antibodies but no longer have the virus (due to the lag between the disappearance of a virus and the subsequent disappearance of its associated antibodies from the bloodstream).[8] These two conditions, although possible, are so rare that an individual who has the HIV test and is interpreting the test results need not even consider them.

The test generally used for HIV antibody testing—the ELISA test—is not 100% accurate. Because of the biochemical nature of the test itself and possible human error in conducting the test, occasional misspecification occurs. As discussed in Chapter 4, there is a very small chance (0.1%, or less than 1 in 1,000) that the ELISA test will indicate that someone has HIV antibodies when he or she does not (false positive) or that someone does not have HIV antibodies when she or he does (false negative). All positive ELISA tests are checked with another test, the Western blot, to reduce even further the likelihood of a false positive. Consequently, the chance of a false-positive test is now extremely small. Negative ELISA tests are not reconfirmed; consequently, the chance of a false-negative test remains the same— that is, very small.

With this background, we are in a better position to answer the question, "Will the HIV test provide the definitive answer to HIV risk?" The answer is, "No, not by itself." The accuracy of the test is very high but not perfect. There is the possibility (although very small) of a false negative. The most troublesome aspect, however, is the window period. This problem can be minimized by a retesting 6 months after the first test. Two HIV-negative tests make the possibility of HIV infection unlikely if, in the intervening 6 months, the individual has not put himself or herself at risk via sex, blood, or birth (see Chapter 7).

The couple mentioned earlier, Thuy and Maricres, can be confident they are HIV negative if they each have two negative HIV tests 6 months apart and if neither has been exposed to HIV during those 6 months in the modes applicable to them; namely, sex or blood. Can Thuy and Maricres, therefore, conclude that their personal risk of HIV infection is zero? The answer is no.

Risk, as we have seen earlier in this chapter, is always relative. Some people have a very low risk of HIV and others have a high risk. Someone who is not and never has been sexually active may be at no risk for HIV through sexual relations, but this same person could be at risk through accidental contact with infected blood. The HIV test provides valuable information about HIV infection in the past; it does not tell about current or future HIV risk. In reality, we are all at some HIV risk unless we live in a protected cell without any contact with other humans.

The HIV test itself does not tell us if we are at risk for HIV. Only an analysis of modes of HIV transmission and their relationship to our lives will give us this information. The HIV test context provides an excellent opportunity for many people to learn about modes of transmission and to consider their relative risks. A very important aspect of this context is the counseling that occurs before the test and when the test results are available.

Testing Context: HIV Counseling as Part of the HIV Test

Taking an HIV test is not like taking most other tests. The consequences of an HIV test are not the same as, for example, those of an allergy test. A positive HIV test has implications for nearly every aspect of a person's life and raises the likely specter of illness and possibly early death. A negative test, on the other hand, can be misinterpreted as a certification of no HIV risk and therefore as a "green light" for behaviors that could put someone at risk of HIV. Both results provide an opportunity for personal risk analysis and the possibility for changes in personal attitudes and behaviors (although, as we will see in Chapter 9, the move from knowledge to attitude and behavior changes is not an automatic one).

Because those taking an HIV test may misunderstand the test and its results, and because there is an opportunity for education, counseling before taking the test and during the explanation of the results has become the accepted procedure. During the pretest counseling, trained health workers explain what the HIV test does and does not do, what the implications of the different test results are, and what type of reporting and recording will occur (confidential or anonymous). In addition, the counselor explores possible modes of HIV transmission with the individual and assists in personal risk assessment.

The other part of the counseling—the time when the test results are shared and discussed—is as important as the pretest counseling. In the not-so-distant past, there were stories of people receiving their test results via mail or even from a message on their home answering machine. In the cases of positive HIV test results, conveying antibody test outcomes in these ways is unethical and cruel, given the implications of a positive result. Even with a negative test result, impersonal communication wastes an excellent opportunity to instruct people about HIV/AIDS and about the continuing need to protect themselves. Instead, face-to-face posttest counseling is the standard mode, done by counselors who are prepared to help people accept and begin to deal with a positive HIV test result or to caution those with a negative test result about such issues as the window period and the need for continued attention to HIV risk protection.

Testing Options

Rapid Testing

Since late 2002, rapid HIV testing methods have been available. These tests involve a saliva swab from the mouth or a blood drop from the finger. In the case of the blood test, a drop of blood is collected and transferred to a vial containing a developing solution. A test strip is then inserted into the vial. If HIV-1 antibodies are present, the test strip will show two reddish purple lines in a small window on the strip. The test has a 99.6% accuracy rate, but the results of a positive test need to be confirmed with follow-up tests, just as occurs with the ELISA test. The oral test works in a similar fashion, using saliva placed in a special solution (Figure 8-1).

Initially, the greater cost of these tests, relative to the commonly used ELISA test, limited their widespread use. Now, however, the cost has been greatly reduced, and it is common to have free HIV testing available that uses rapid testing.

From a public health standpoint, the advantages of rapid testing are significant enough that many organizations and groups advocated for greater use of these tests. In particular, a rapid test does not require someone to return in several days or a week for test results; instead, the test and its results are completed in one session. A major challenge with the common ELISA testing is the requirement for those being tested to

Figure 8-1   An OraQuick test instrument. (Courtesy of OraSure Technologies, Inc.)

return for the results after the test has been sent to the laboratory for processing and reporting. At some testing sites, half or more of those tested never returned for their results. With a rapid test, the test is completed and its results are known and shared in 20 minutes. Pretest and posttest counseling are still important but occur in a more condensed fashion. In the cases where the test is positive, the counselor explains that a confirmatory test will be needed but also begins discussing, to a depth appropriate for each individual, the implications of his or her likely new HIV status.

Confidentiality Versus Anonymity of Test Results

There is one other important consideration of testing: Are the test results confidential or anonymous? With *confidential* testing, a record of the test result linked to an individual is maintained. With *anonymous* testing, however, no link ever joins a test result and a person's name. When performing an ELISA test, for the purpose of test laboratory identification, numbers or nonsense names can be temporarily attached to HIV test blood and the associated test result; only the individual whose blood is being tested knows the number or nonsense name. For example, those being tested are given a half of a sheet with the same unique number on both halves; the other half sheet is attached to the blood vial. To retrieve the test result, counselors need both halves of the original sheet. If someone loses the half sheet, the results cannot be claimed and the procedure must be done again. In the case of a quick test, no number or nonsense name is needed because the test and result are done in the same session.

In the first two decades of the AIDS epidemic, anonymous testing was generally favored over confidential testing. This choice related to the unauthorized disclosure of some early test results and to the discrimination that individuals identified as HIV positive frequently experienced, sometimes including violence. The situation has changed, in part because of less discrimination, stronger laws, major changes in the course of HIV progression and treatment, and greater pressure from many in the public health community to have names reporting, along with adequate safeguards, so that follow-up and contact tracking would be possible on a larger scale. In most U.S. states at this point, medical personnel are required to report anyone with a positive HIV test to the state health department. However, even in states with names reporting, some of those tested have found ways to keep the information to themselves. As one staff member at New York's large Gay Men's Health Crisis testing clinic said, "We have a lot of George Washingtons."

If a choice can be made, anonymous testing is still preferable to some people. When only confidential testing is available, the person being tested should understand the ways in which confidentiality will be guaranteed and the extent to which results will be reported to others, including local and state health officials. It is important that the person being tested feels comfortable with these procedures and practices before proceeding with the test, because once a positive test result occurs, medical personnel may have no choice except to report the result.

At the beginning of this chapter, we presented a model for individual decision making and action involving four steps: knowledge, attitude, intention, and behavior. This chapter focused on the knowledge and attitude steps with regard to assessments of HIV risk. Based on the outcomes of these assessments, individuals form intentions to take actions—or to take no action—and then to maintain or change their behaviors related to HIV risk. As we shall see in the next chapter, knowledge does not automatically result in behavior changes. Preventing HIV transmission and AIDS requires much more than knowledge.

Notes

1. A good introduction to this area is the book by Dan Ariely, *Predictably Irrational: The Hidden Forces that Shape Our Decisions* (New York: Harper Collins; 2008).

2. This example, as well as most judgment heuristics described for the subjective probability model, are drawn from A. Tversky and D. Kahneman, Judgment Under Uncertainty: Heuristics and Biases. *Science* 185: 1124–1131 (1974).

3. The frequencies from the U.S. Department of Labor, Bureau of Labor Statistics, for May 2008, the most recent available, are as follows: salespersons 4,426,280; farmers 985,900; physicians/surgeons 661,400; librarians 159,900; and pilots 76,800.

4. The Tversky and Kahneman article (ibid) discusses other factors that contribute to the availability heuristic; familiarity and salience are particularly important ones for our more limited discussion here.

5. Ibid.

6. Two examples of other relevant events: the concentration of HIV in Monica's blood at the time of the bleeding (we know that concentrations vary over time) and the extent to which blood from the cut on Monica's arm has dried by the time Brad touches it (we know that HIV dies quickly when exposed to the drying effects of air).

7. These other factors include perceived control over the risk, fear, newness of the risk, and potential for catastrophic outcomes.

8. An HIV-infected mother will pass HIV antibodies to her newborn but may or may not pass HIV. Consequently, her newborn will test HIV positive but may not, in fact, have HIV. In the absence of HIV, the antibodies may not leave the infant's system for up to 18 months. If HIV has also been passed, antibodies will continue to be produced, and the child will continue to test HIV positive.

http://biology.jbpub.com/fan/aids/6e/

Connect to this book's website: http://biology.jbpub.com/fan/aids/6e/. The site features summaries of the main points from each chapter, links to important AIDS-related websites, and short-answer-style review questions for each chapter.

CHAPTER 9
Prevention of
AIDS

The key to preventing AIDS is to stop the transmission of the HIV virus before it enters the human body. No vaccine has yet been developed to produce immunity or prevent the disease, and researchers are not hopeful about the development of one in the foreseeable future. Our focus on preventing AIDS, therefore, needs to be on preventing HIV infection. Because we know how HIV is transmitted, we also know how HIV can be blocked from passing from one infected person to another (see Chapter 7). This knowledge needs to be disseminated to anyone who is at risk for HIV/AIDS. In this chapter, we review general principles of disease prevention and then apply them to the case of preventing infection with HIV. We conclude with two examples of successful HIV/AIDS prevention programs.

Disease Prevention and Health Promotion

The prevention of AIDS appears to be simple. First, we give the information about stopping HIV transmission to people who are at potential risk, and, second, they act on this information. The information includes suggestions to avoid certain very risky sexual practices, to use condoms, to stop sharing needles, and to avoid direct contact with certain human body fluids. (Chapter 7 contains the specific details.) Once informed, everyone would follow HIV/AIDS prevention measures, and the spread of the virus would be halted, preventing AIDS.

Unfortunately, the task is not at all this simple. Disseminating HIV/AIDS knowledge is different from changing HIV/AIDS attitudes and intentions, which is different again from changing behaviors that put one at risk for HIV/AIDS. Health promotion researchers have discovered that people generally are resistant to changing both their attitudes and behaviors. Why do people not change their health-threatening behaviors when the health risk is obvious and when the means of prevention is clear and effective? Perhaps people do not understand the risk and the way to prevent it. Research has shown, however, that even when people clearly understand a health risk and the means of prevention, they still are resistant to change. Perhaps people do not want to change. Additional research has demonstrated that even in cases where people understood the risk, accepted the prevention method, and reported that they wanted to and would change, subsequent behavior did not always change.

Health promotion and disease prevention researchers have organized their findings into several models, which highlight important factors explaining why people are resistant to changing their health-related behaviors. In the sections that follow, we turn first to a description of these models and their important concepts and then to a discussion of principles largely derived from these models, which give us guidelines about how to plan and implement effective disease prevention programs. The task is not an easy one, as many have discovered with other disease prevention programs. But, as we will show with several AIDS prevention program examples, when prevention programs are done well, they can be effective and can change behaviors related to HIV transmission.

Models of Health Behavior Change

Three models related to health behavior change are particularly useful in understanding HIV/AIDS prevention. These models incorporate concepts and ideas from a long history of research on measuring, establishing, and changing attitudes and behaviors. Each of the three health behavior change models is described below.

Health Belief Model

This model is the oldest of the health behavior change models and grew out of researchers' initial investigations of the reasons for the widespread failure of people to take actions to prevent asymptomatic diseases. (*Asymptomatic diseases* are those such as lung cancer, in which no symptoms appear until the very end of the disease, when death is likely and nothing can be done to reverse the damage. Symptomatic diseases—a cold, for example—have immediate harmful or unpleasant effects that we are aware of; these noticeable effects make it more likely that we will take action.) The health belief model identifies three main variables to explain the absence of action. All three variables focus on an individual's perception of different aspects of a health-threatening situation.

The first variable is the person's perceived susceptibility to a health threat. Someone who does not see himself or herself as "at risk" will not change a health-related attitude or practice. In Chapter 8, we discussed why many teenagers do not see themselves as being at risk for many diseases, including AIDS. Their view of their own susceptibility— their perceived susceptibility—to certain diseases is the critical factor, not the objective measure of their susceptibility. If a person believes that he or she is not susceptible to a disease, no matter what the actual degree of susceptibility may be, that person will not begin the process to protect himself or herself from contracting the disease.

The second important variable is an individual's assessment of the severity of the threat. A person may acknowledge that he or she is susceptible to a particular disease but then rate the severity of the threat as too low to worry about. If this person judges the severity to be low, there is not much incentive to take protective action. Again, as was the case with the variable of perceived susceptibility, it is the person's subjective assessment that is important, not the objective measure. We know that AIDS is a deadly disease in almost all cases, a disease with the most severe threat possible. If a person does not know, understand, or accept this, however, that person will continue to underrate the severity of AIDS, and this is one factor that may explain his or her lack of action.

The third variable is a person's evaluation of the effectiveness of the recommended health-promoting or illness-preventing action. If the action is clearly effective, it is easier for a person to decide to undertake the action, or, conversely, it is more difficult for a person to make excuses to avoid taking the action. Seatbelt use in automobiles is a good example. As the data have become conclusive that seatbelt use saves lives, more drivers have become regular users of seatbelts. The HIV/AIDS case provides a different sort of example, one with a more confused outcome. Condoms are promoted as one way to minimize the risk of contracting HIV during sex. Properly used, condoms provide a very large measure of protection but not 100% protection. Because condoms are not completely effective, some people have decided that they are not an effective AIDS prevention device. Others, however, see condoms as a generally effective way to prevent the spread of HIV. Those who underrate or denigrate the value of condoms in preventing the spread of AIDS feel little or no pressure to use them, which may explain why some people do not use condoms.

Health Decision Model

This model is a more recent reformulation of the health belief model that incorporates variables beyond those related to the individual's views. Decisions about health actions are often made in the context of other people, with other people's views either explicitly considered, in cases where two people must jointly take an action (such as using a condom to prevent the spread of HIV), or implicitly considered, in cases where there is an individual action, but it is taken in a social context (such as a decision to stop smoking). The social context is especially important in the case of AIDS. Condom use, for instance, is not an individual decision only; it is a decision made by two individuals. When a man and woman decide to use a condom during sex, they make a joint decision, even though it is the man who wears the condom. If the woman favors the use of the condom and the man does not, or vice versa, the conflict needs to be jointly resolved.

The health decision model focuses attention on the social variables of experience, knowledge, and interaction, in addition to the three health belief model variables of perceived severity, susceptibility, and evaluation of action. This expanded model acknowledges that decisions to change health-related attitudes and behaviors are made with some attention to our past experiences with other people who are important to us, our knowledge of others' views and opinions, and our current interactions with others.

The health decision model gives us additional insights into why someone may or may not follow a recommended action to prevent the spread of HIV. First, we need to look at the larger social context in which the person lives. For example, in the case of condom use, we need to consider cultural values related to condoms to understand why using condoms might be easy or difficult for people from different cultural groups. In Hispanic culture, for instance, condoms have a number of negative associations (the Catholic Church is against them; prostitutes are associated with them), which present additional obstacles for a Hispanic male who is considering protection against HIV.

This example also illustrates the importance of experience and knowledge, the other important social variables in the health decision model. If a Hispanic male has not seen or heard about condoms before attending an AIDS prevention program or has heard only negative things about condoms, it will be even more difficult to convince him to use condoms to prevent HIV infection. He may know about AIDS, see himself as highly susceptible, see AIDS as very serious, and understand that condoms help prevent AIDS, but he still may not take action because of his past knowledge and experience (or lack thereof).

Precaution-Adoption Process Model

Finally, we look at one additional model that has been proposed to help explain why health-related attitudes and behaviors are not easy to change. This model—the precaution-adoption process model—focuses on the process of change rather than on particular variables, as the previous two models do. Both the health belief model and the health decision model are static and linear in their view of the health attitude and behavior change process. That is, these models assume that a person moves very logically from A to B to C to D and reaches a decision. For the case of HIV/AIDS risk, for example, a person would assess personal susceptibility to AIDS, severity of the AIDS threat, and the efficacy of the recommended action (say, condom use), using past experience and knowledge, particularly from discussions and interactions with friends and partners, as important bases to make these assessments. From this assessment, the person makes a decision whether to use a condom.

Laying the situation out in this way highlights how artificial such a process would be for most people. Instead of a smooth step-by-step decision process, most of us go through a much more complex decision sequence, making one step forward and a half step back, waiting for a while, then taking another step forward, eventually reaching—or backing into—a decision. Some researchers recognized that, for most of us, decision making is dynamic and fluid; different factors come into play in different ways at different times. In the case of AIDS prevention behaviors, for example, certain behaviors (say, condom use) require that two people make a joint decision. The important decision factors may be weighed differently for each of them, and as they move toward a joint decision, the weights of factors may change in different ways, further complicating a final decision and action.

The precaution-adoption process model proposes that people go through five stages in deciding to make behavior changes:

1. Awareness or knowledge of a risk or threat
2. Acknowledgment of a significant risk to some group of people
3. Acknowledgment of a significant risk to oneself
4. Decision to take action to reduce the risk
5. Initiation of the behavior

Movement through the stages can be forward or backward as a person's emotions, values, experiences, knowledge, intentions, actions, and social context change over time.

These three models are not in competition with one another but instead are complementary and provide different perspectives on the health attitude and behavior change situation. The more varied our perspectives, the more likely we are to understand obstacles to change and, as important, factors to facilitate change. This is the topic to which we now turn.

Principles of Health Behavior Change

Drawing upon the three models just described and other research as well, health behavior change researchers and program personnel develop programs to foster health-related attitude and behavior changes. The general concepts and principles that guide them are summarized here. This set of seven health behavior change principles[1] incorporates factors from the three models and adds several new concepts.

The Cognitive Principle

Correct and relevant knowledge must be conveyed to those whom you want to make changes. People need to know the "facts" about HIV: its risks, its spread, and its prevention. Sometimes this involves conveying new information; at other times it involves correcting misinformation. In the case of AIDS, the cognitive information elements of an AIDS prevention message could include what HIV is, how it is and is not transmitted, how likely people are to become infected, what the consequences of AIDS are, and how AIDS can be prevented.

The cognitive "facts" about HIV/AIDS are numerous (as demonstrated by the length of the earlier chapters in this book). How can we possibly have all the facts in a brief AIDS prevention message? We cannot include, and do not need to include, all the AIDS facts in every message. The challenge is to decide which facts are critical to convey and how they can be most effectively communicated. Generally, this information needs to be simple and in language that is appropriate for the intended target audience. Some of the principles discussed later will help us decide which HIV/AIDS facts need to be conveyed and how these facts can be communicated best. However, even though knowledge about HIV/AIDS facts is necessary by itself, it is not sufficient for someone to change. We can barrage someone with information, but unless we communicate appropriately, the message will not have its intended effect. The other principles discussed here must also be considered in developing an AIDS prevention message.

The Emotional Principle

Change is facilitated when a connection is made with a person's emotions. Rather than simply conveying information to a person, we should create an emotional "hook" with the information. Love or romantic emotions are one choice; think of the number of advertisements that use this emotion as their hook. Positive emotions generally are better hooks than negative emotions, but fear is one emotion that can be effectively used in health behavior change communications. We have to be careful with this particular emotion, however, because if we generate too much fear, our message may be missed or even backfire. An example of this was the early "stop smoking" campaigns, which used such vivid and gory pictures that some smokers literally turned away from the message and never received it, and other smokers either rejected the information outright or decided it was too late to make a change—and kept smoking. To know how much fear is appropriate or to learn which of the other emotions would be best to tap, we need to know and understand well the intended audience for our message. Aspects of several of the other principles presented here can assist in this task.

The Behavioral Principle

A successful program recommends specific behaviors that a person should adopt. The more specific the behaviors recommended, the better. For example, a vague communication to "avoid getting AIDS" does not describe the specific behaviors that should be followed. The message "use condoms during sex" is much more specific and therefore more likely to be followed. However, even this message may not be specific enough for certain target audiences, such as those who have had no experience with condoms. In this case, the message ought to include explicit instructions on how to use a condom, ideally with hands-on experience.

An old AIDS prevention education campaign offers an example of the improper use of this principle. A national advertising firm created an audiovisual AIDS prevention message. The audio portion advised listeners to take precautions to avoid AIDS; the accompanying video component showed a man putting on a sock. The idea was that putting on the sock symbolized putting on a condom. Alas, the behavioral message—both audio and video—was too vague to be useful. Indeed, many viewers, not surprisingly, missed the connection between the sock and the condom altogether.

The Interpersonal Principle

An effective message should consider the immediate social network of the target person. Individuals do not exist in isolation; they are part of social groups, both small and large. Social groups exert influence on their members; we cannot ignore this influence and still expect an effective health behavior change to occur. Particularly in the case of AIDS, in which the virus is transmitted between individuals, we have to take into account the social network of those we hope to influence with HIV/AIDS prevention messages.

Consider the case of a sexually active heterosexual woman who is our target for a particular AIDS prevention message, namely, to use condoms during sexual relations. Not only must our message convince the woman of the necessity and effectiveness of condom use, but it also must include information about and strategies for negotiating condom use with her male sex partners. We have to convince the woman of the need for protection against AIDS and then train her about ways to convince her sexual partners. This is a very challenging task both for the health promotion campaign and for her. It is a challenging task for the campaign because each man whom the woman encounters will have a different perspective on the AIDS situation and on the need to use condoms. It is a challenge, therefore, to develop a succinct health promotion message that will address the many different partners the woman may have. The challenge for the woman is even greater. Not only will each potential sexual partner be different in his general attitudes about HIV/AIDS and condom use, but each will also vary in his personal mood at the moment when the woman must initiate her AIDS prevention communication.

This example shows how difficult it is to initiate and achieve AIDS prevention measures. We have to consider not only the individual but also the other people who relate to the individual, either on a one-to-one basis, such as potential sexual partners, or on a more general basis, such as the members of our family or of a social group that is particularly important to us. In the cases of our family or our social groups, we need to be aware of the social norms that govern sexual behavior and related AIDS prevention strategies. Consider the case of a homosexual man whose friends do not like to use condoms. We would probably be unsuccessful in our AIDS prevention campaign to him if we focused solely on the risks of HIV/AIDS and why he should use condoms. We would also have to address the social norm operating in his group of friends—the negative perception of condom use. We would first have to know more about why his friends have this negative view (the feel of sex using a condom? the inconvenience of using a condom? the associations with using a condom, such as that it could imply that someone does not trust his partner?) and then develop our message to address the specific concerns.

The Social Ecological Principle

General social and cultural issues specifically relevant to the target individuals should be considered in developing effective HIV prevention programs. The interpersonal principle recognizes the importance of those people who surround the target individual in achieving change in health attitudes and behaviors; we could consider this the inner circle of social influence. The social ecological principle recognizes the importance of circles of individuals beyond this inner circle and the interrelatedness of all these circles of social influence that surround our target individual. In particular, the social ecological principle addresses the importance of the social and cultural dimensions.

All of us are part of different social and cultural groups. These groups have norms—or standards of behavior—about many activities, including sexual activity and drug use, the two behaviors in which we are most interested as AIDS prevention specialists. To be effective HIV/AIDS educators, we must be aware of the particular social and cultural norms related to AIDS of the groups to which our target individual belongs and of how these norms are similar to or in conflict with AIDS prevention guidelines. In addition, the HIV/AIDS educator should recognize the benefits—and sometimes the necessity—of establishing new social norms that promote AIDS prevention. When the social atmosphere promotes change toward AIDS prevention, rather than discouraging it, the individual finds it easier to initiate and follow through on HIV/AIDS preventing activities.

The situation with HIV/AIDS education and drug users presents a good example of the need to attend to social and cultural norms. If we consider only the cognitive principle in developing an AIDS prevention message related to drug use, the message is straightforward: Use new needles for intravenous drugs, do not share needles, and clean needles with bleach if you must share them. In the United States, this message has generated controversy, not because the facts are wrong but because the content clashes with general social norms. The drug user or those in his or her inner circle may accept the content of the message, but others in the outer circles of the general population resist the message because they disagree with any recognition and tacit acceptance that someone is using drugs. For example, some of those in the outer circle say that providing people with clean needles or telling people to clean their used needles legitimizes drug use. They prefer the message "don't use drugs." Unfortunately for the drug user, it is not a simple matter of stopping drug use; a physical addiction cannot be easily ignored. In addition, the drug user is part of a subculture in which drug use is accepted and the social norm is to continue the behavior.

The approach that HIV/AIDS educators have generally taken to solve this conflict is to develop programs for particular groups of drug users, using their social norms as the basis and starting point for AIDS prevention actions. Because the general population is not the target population, AIDS educators working with drug users usually keep a low profile in the general community to minimize the conflict with social norms of people who do not admit to or recognize the realities of drug use.

The social ecological principle also directs our attention to the necessity of having role models or leaders as part of the health campaign. Our message can be much more effective if it comes from a respected leader or acknowledged expert rather than from an unknown announcer. If the model or leader is viewed as a social norm setter, his or her message will have more effect on the targeted listener because that individual represents a bridge between the individual's inner and outer circles of social influence. The challenge is to pick the most appropriate role model for the individuals we want to reach. Different models are required for different individuals, even different models for the same individuals as the situation changes. For instance, we may need one type of role model when we are first trying to convince someone to use condoms and then a different type of model when we provide more detailed instruction about the use of condoms, about how to negotiate condom use with a sexual partner, or about maintaining condom use over time.

The Structural Principle

An effective prevention program considers the laws, technology, and physical settings relevant to the target individuals. The context in which an individual lives involves not only social and cultural dimensions, but also laws, technology, and physical settings. These latter aspects are considered the structural elements, and they can complement or conflict with the social and cultural norms. For instance, laws are the codification of some social and cultural norms, but frequently a gap exists between current laws and current norms, especially regarding sensitive social issues such as AIDS. The HIV/AIDS prevention specialist, therefore, must be aware of these realities when he or she develops an AIDS prevention campaign. In important ways, the structural elements can support or hinder health promotion messages and programs.

A good example of the power of structural elements in health promotion campaigns is provided by the case of efforts to decrease smoking. The data (that is, the cognitive aspects) are clear about the harmful effects of smoking on the smoker and on those in his or her environment. Smokers generally were aware of these deleterious effects, but this knowledge alone was not usually sufficient for them to stop smoking. In addition, certain groups (especially young women) were continuing to smoke. After the lackluster success of antismoking campaigns aimed at the general population, antismoking advocates changed their strategy to focus on structural changes. Now, laws regulate smoking and physical barriers to smoking, such as the nonsmoking rule on all airlines and the increasing number of nonsmoking rules for restaurants and other public places.

In the case of AIDS, structural elements are equally useful. For example, we encourage people who are at high risk for AIDS to be tested for HIV. Before we promote this activity, however, we need enough testing sites in place with enough well-trained counselors to carry through on the benefits of testing. Likewise, laws have to be in place providing for anonymous testing, or at least for confidential testing to minimize or eliminate potential harmful consequences of the release of results.

Another good example of the importance of structural elements relates to the issue of condoms. Condoms are widely advocated as good protection against HIV transmission. Consequently, it has been important to attend to the quality of condoms. Tests are now routinely conducted on the structural integrity of condoms, and the results are disseminated to increase the production and availability of only the highest quality condoms.

A different sort of attention has been given to other physical aspects of condoms. Condoms for males are available in a variety of styles, sizes, colors, and even tastes to increase their attractiveness to potential users. A female condom (that is, a plastic sack-like device inserted into the vagina) has also been developed, although it is less readily available and more expensive than male condoms. There are some stores in some major cities around the world devoted largely to condoms. (Interestingly and importantly, these stores also create a social setting—think back to the social ecological principle—in which condoms are the norm and the focus of attention; this helps to reorient social norms surrounding condoms, changing them from a taboo topic to a common—or even humorous—one.)

The Scientific Principle

Prevention programs should be developed and evaluated using the scientific method. When developing an HIV/AIDS prevention program, scientific research methods can help us determine the best components for the program. For example, focusing on a particular target group, we can determine deficits in AIDS knowledge or current AIDS-related social norms, then use this information to create a program for the group. Once developed, we can use scientific methods to test the program components in a formative way, to be sure the components are working properly, and then in a summative way, to determine the overall effects and effectiveness of the program. In these ways, HIV/AIDS prevention programs are put to the test using the scientific method—looking systematically and in an unbiased manner at causes and effects. The effects of a program are measured, using objective assessments, against a control or comparison condition to determine whether the program outcomes are occurring as hypothesized and also to uncover other unexpected outcomes, either positive or negative.

In a typical scientific evaluation, quantitative indices of the program's goals are developed. Two identical groups are formed from members of the target audience. Baseline measures are made on members of the groups before the program begins. Then, one of the groups receives the program and the other, serving as the control, does not. (On occasion, the control group receives a limited version of the program or a different program, or receives the program later, after it has served as the control.) After the program is concluded, both groups are measured again to detect changes attributable to the AIDS prevention program. In addition to quantitative measures, interviews are frequently conducted with those in the program to obtain information about their experiences—both positive and negative—with the program components. Sometimes, we inadvertently cause some changes—sometimes for the better and sometimes for the worse—when we implement a health promotion program, and only the participants can tell us about these.

Increasingly, program evaluation is becoming a regular part of AIDS prevention programs. At the outset of the AIDS crisis, there was a tendency to rush to implement any type of program, without sufficient attention to its consequences. The lack of changes resulting from some of the early HIV/AIDS education campaigns caused those involved in AIDS prevention to take a harder look at the concrete results of their efforts and to employ more rigorous scientific methods. The two examples that follow are good examples of how AIDS education programs can be evaluated.

Examples of HIV/AIDS Prevention Programs

The two AIDS prevention programs described next incorporate aspects of many of the seven principles of health behavior change. The two programs are very different: The first program focuses on young gay and bisexual men and the second on Mexican migrant farm workers.

AIDS Prevention Among Young Gay and Bisexual Men[2]

This program, called the Mpowerment Project, focused on young gay and bisexual men, from ages 18 to 29 years. The project, which was developed and implemented by young gay men, aimed to reduce unsafe sexual practices, specifically unprotected anal intercourse, among these young men. The project involved four related activities that occurred in a physical space that, ideally, was dedicated to and associated with many of the project's activities.

1. Formal Outreach. Small teams of young gay men went to places where other young gay men congregated. The team members distributed informational materials developed specifically for these men related to HIV/AIDS, handed out condoms, and engaged the men in general discussions about safer sexual practices. In addition, the teams of men working on the project developed special social events to involve young men in fun activities, such as dances and picnics, where they again promoted safer sexual behavior.

2. M-Groups. These were small 8- to 10-person groups focusing on reasons for unsafe sex among the men. Led by a trained peer, the one-time discussions lasted for 2 to 3 hours and covered many topics, from the ease or difficulty of condom use to communication between sexual partners about sexual practices. In addition, the sessions involved skill-building and training on condom use and how to negotiate safer sex with a partner. Participants received free condoms and lubricants and were encouraged to, and trained for, informal outreach to their friends and peers.

3. Informal Outreach. These were the discussions and conversations that program participants initiated on their own, using the project materials, to inform their friends about and promote safe sexual practices.

4. Ongoing Publicity Campaign. To advertise the project, there were articles and advertisements in local publications directed toward the local gay male community.

Referring back to the seven principles presented earlier, it is clear that the Mpowerment Project involved a number of them. The information that was distributed by the men, along with the content of the discussions, involved the cognitive principle. The young gay men created the material themselves, so it had a format and content that was appropriate and appealing to this particular group. The emotional principle was also involved, in a positive way: The project involved enjoyable, fun activities. The focus of the project on condom use clearly specified a behavioral option that the young men could use to protect themselves from HIV. The interpersonal principle was also involved in the networks that were formed among the project participants and the opportunities these gave the young men. The benefits of the social ecology principle were employed in the way that the culture of young gay and bisexual men was the anchor for and framework that surrounded all the components of the project. The structural principle was not a part of the intervention (no activities focused on changing laws or policies, for example).

Lastly, the scientific principle was a central part of the program. To assess the effectiveness of the Mpowerment Project, two similar communities were used. The Mpowerment Project was implemented in one community but not in the other. Before the project started, large groups of young gay men from both communities were surveyed and asked a number of questions, in particular about unprotected anal intercourse with various partners. In the community that had the Mpowerment Project, the percentage of young men reporting any type of unprotected anal sex decreased from 41% to 30% (a 27% drop from baseline); those reporting this behavior with nonprimary partners decreased from 20.2% to 11.1% (a 45% drop from baseline); and those reporting this behavior with boyfriends decreased from 58.9% to 44.7% (a 24% drop from baseline). These changes were analyzed and found to be statistically significant. The young men in the comparison community, which had not experienced the project, showed no significant changes. This allowed the project developers and researchers to conclude that the Mpowerment Project caused a meaningful change in the young gay and bisexual men's behavior, reducing their HIV risk.

AIDS Prevention Among Mexican Migrant Farm Workers[3]

This program was conducted in Southern California in the early 1990s. It involved about 300 male farm workers who traveled to the United States to pick crops in the agriculture fields. Migrant workers in this area are at risk for AIDS because prostitutes are regularly brought through the camps where they live. These prostitutes are sometimes infected with HIV, and because of the large number and rapidity of sexual contacts these women have with farm workers and others, a risk exists for the spread of HIV. A program was developed to instruct the farm workers about AIDS and about the need to use condoms with prostitutes. The main component of the program was a *fotonovela*—an eight-page photo storybook with pictures and captions. Figure 9-1 provides a sample of some of the fotonovela panels. The story tells of three farm workers who meet prostitutes and through the prostitutes learn about the need to use condoms. In addition, condoms were provided, as were instructions on how to use them. (The program had other components, but we will not focus on those aspects here.) The program was developed with farm worker input about the best approach to take with such culturally sensitive issues as sex and condoms. Farm workers also served as models for the photos.

By looking closer at the program components, we can see that all of the health behavior change principles are involved in this intervention: the cognitive, emotional, behavioral, interpersonal, social ecological, structural, and scientific principles. The knowledge and facts about AIDS conveyed by the fotonovela involve the cognitive principle. Using the cognitive principle, knowledge was conveyed in an unusual way: not by giving a list of do's and don'ts but instead by telling a story that had several plot lines, one of which involved one of the men learning how to protect himself. Through

Marco, Sergio and Victor—three men who leave their town in search of opportunities, they confront danger!

A condom? Why?

Condoms give protection to both of us!

Protection?

Figure 9-1   Sample set of photos reproduced from the fotonovela, "Tres Hombres Sin Fronteras" (Three Men Without Borders), an AIDS Prevention Photo Booklet for Mexican Migrant Farm Workers. (Reproduced from Novela Health Foundation [1989]. *Tres Hombres Sin Fronteras.* Granger, WA: Novela Health Foundation.)

the story, readers learn how condoms can prevent HIV transmission and AIDS. Several of the emotion-filled story lines in the fotonovela use the emotional principle. The explicit instruction about how to use a condom relates to the behavioral principle.

The social ecological principle is involved in the use of farm worker and prostitute models in the photos with which the men could identify. (Preliminary research, for instance, showed that the prostitute who advocated condom use was viewed by the men as a "higher class" prostitute, exactly the type, they said, who would believably

Yes, protection against diseases like gonorrhea, syphilis, even AIDS!

Because when you get AIDS, there isn't a cure for that.

So then?

OK!

Figure 9-1 Continued

advocate condom use.) The social ecological principle was also involved in the social and cultural considerations taken into account for the format of the AIDS materials; a fotonovela is a format with which Mexican men are very familiar and one that is at an appropriate literacy level. The structural principle was also involved: The condoms given to the men were specially selected not only for high quality but also for a degree of lubrication that these men were known to prefer. Finally, the scientific principle was involved in the development of the materials and in the design of the study that assessed the effects of the materials.

The scientific study of the program's outcomes involved using small groups of farm workers, all of whom were tested about AIDS-related knowledge, attitudes, and behaviors before any program began—the pretest (baseline) measure. Then, about two-thirds of the men (the test group) received the materials, but the other men did not and served as a control to assess the effects of the materials over about a month's time. After this, all the men were again tested—the posttest measure—and the fotonovela and other materials were given to the one-third of the men who had not yet received them.

In brief, the results showed that all these men knew many facts about AIDS even before participating in the program. The fotonovela materials resulted in small but significant changes in knowledge (comparing the "pretest" and "posttest" measures of the men in the test and control groups). The most significant changes were in reported condom use with a prostitute—the target behavior. Among the men who had received the fotonovela and who had the opportunity to use a condom with a prostitute, all but one had used condoms. Among the control group of men who had not received the fotonovela and who had the opportunity to use a condom with a prostitute, none had used a condom. An important lesson from this study is that AIDS prevention materials must be very carefully targeted, taking into consideration all the principles we have reviewed.

Although both of these AIDS prevention programs produced positive changes, neither resulted in perfect compliance with AIDS prevention methods. In the study with young gay and bisexual men, some of the young men were still engaging in risky sexual behaviors. In the study on migrant workers, men at one of the camps did not make the desired behavior changes. These less-than-perfect outcomes remind us that the path from new health-related knowledge to changes in attitudes, intentions, and behaviors is a challenging one. With the help of the scientific principle, we can identify when and why we are successful and build on these factors, and we can identify when and why we are not successful and then change these factors to make the program better.

In this chapter, we focused on the challenge of preventing HIV transmission and AIDS. As the two examples showed, successful AIDS prevention involves careful attention to the cognitive, emotional, behavioral, interpersonal, and social ecological situation of the targeted individuals, as well as attention to the structural aspects of the context in which the targeted individuals live. After we have considered all these aspects, we are then ready to develop our AIDS prevention program. When we implement the program, we need to carefully assess its effectiveness with a rigorous scientific approach. Our scientific assessment must be a regular part of the program because people and situations change, and we need to be aware of those changes.

Notes

1. This set is an expansion of a related formulation in L. Liskin, C. A. Church, P. T. Piotrow, and J. A. Harris, AIDS Education: A Beginning. *Population Reports*, September 1989 (Series L): 8.

2. Full details are in S. M. Kegeles, R. B. Hays, and T. J. Coates, The Mpowerment Project: A Community-Level HIV Prevention Intervention for Young Gay Men. *Am. J. Public Health* 86 (8): 1129–1136 (1996).

3. More details are in *AIDS Crossing Borders: The Spread of HIV Among Migrant Latinos*, S. I. Mishra, R. F. Conner, and J. R. Magaña (eds.). Boulder: Westview Press; 1996.
4. These materials were developed by the Novela Health Foundation, supported by a grant from the California Community Foundation; used with permission. The study of the materials' effectiveness was also supported in part by the California Community Foundation, along with the support of the American Foundation for AIDS Research.

http://biology.jbpub.com/fan/aids/6e/

Connect to this book's website: http://biology.jbpub.com/fan/aids/6e/. The site features summaries of the main points from each chapter, links to important AIDS-related websites, and short-answer-style review questions for each chapter.

CHAPTER 10

Living with AIDS: Human Dimensions

Theoretical Perspectives from Social Psychology
Role Theories
Cognitive Theories

Human Dimensions of HIV/AIDS
Confronting the News of Infection
Accepting the Reality of Infection
Opportunities and Challenges of Drug Therapies

All of us live with AIDS. Some of us have HIV, others have full-blown AIDS, and still others are HIV negative. Whatever our HIV status, however, all of us, as members of our society and the interconnected world, are living with the realities of HIV/AIDS, either indirectly or directly.

Some people are more directly affected. We frequently hear about "PWAs," people living with AIDS. These are HIV-positive people living with AIDS day in and day out. Others are living with AIDS almost as directly because their spouse or lover, brother or sister, coworker or neighbor is HIV positive. Other people are living more indirectly with AIDS, even though they may not be aware of it. Because members of every community are infected with HIV, changes have been made in the content of and budgets for our cities' and communities' health, social service, welfare, and educational systems. Laws and policies have been changed because of HIV/AIDS. On a cultural level, HIV/AIDS has brought certain issues out of the shadows, such as homosexuality or "male-to-male sex" (as certain cultures prefer to label it), that have catalyzed discussions and changes about relationships between citizens.

Everyone in U.S. society and in nearly every society around the world, therefore, lives with HIV/AIDS. In this chapter and the next, we analyze two aspects of living with AIDS: the human dimension and the societal dimension. In this chapter on the human dimension, we focus primarily on those who are HIV positive and the challenges they face. In Chapter 11, we will consider some of the important societal effects of HIV/

AIDS. In both chapters, we draw on research from the field of social psychology, which studies the reasons for and effects of human interactions of all types. As the basis for our discussion of the human and societal dimensions of HIV/AIDS, therefore, we need first to understand two important sets of theoretical perspectives from social psychology.

Theoretical Perspectives from Social Psychology

The field of social psychology concentrates on human behavior in groups. Many aspects of our behavior are determined by the direct or indirect influences of others, even some aspects that we may believe as "innate" or "inside" us and therefore beyond the control of others. For example, our self-concept is formed and reformed in interaction with others. Our attitudes and beliefs also are shaped and reshaped through discussion and interchange with other people. Indeed, because we are social animals who live in groups, few aspects of our inner or outer selves are unaffected by other people.

Social psychologists over the past century have developed theories to explain different aspects of people's behaviors in groups. No theory explains all aspects of people's behaviors, but theories taking a similar approach have been useful in explaining different parts of people's behaviors. These similar approaches can be grouped into three general theoretical perspectives.[1] These three perspectives are role theories, learning theories, and cognitive theories. Role theories and cognitive theories are particularly relevant to our understanding of the human and societal dimensions of HIV/AIDS. Consequently, these perspectives are described in more detail in this chapter. For the sake of completeness, learning theories can be briefly described here as those that focus on the relationship between stimuli (such as rewards and punishments) and responses (such as changes in personal behavior). From this perspective, for example, human behavior is viewed as being affected by an exchange of rewards (both concrete ones like money and abstract ones like prestige). Some aspects of the human and societal dimensions of AIDS are illuminated by this theoretical perspective, but it is less relevant to us than are the other two theoretical perspectives from social psychology, to which we now turn.

Role Theories

The basic idea that underlies this set of theoretical perspectives is that the roles we have are significant determinants of the ways we interact with others and the ways others interact with us. Before we progress further, we should raise a caution: The theatrical concept of a role being played on stage is not the correct conception of *role* for this set of theories. A theatrical role is one that is superficially adopted by the actor. Also, the theatrical idea of *role* has the connotation of artificiality; an actor can decide to "play" a role and behave in artificial ways dictated by the role. If you think about really good actors and actresses, however, who go beyond superficiality and artificiality to genuinely "get inside the skin" of characters, you begin to understand the kind of conception of a role that we want to think about here.

Role theorists see *roles* as the fibers of the social network that interlinks all of us. Think of the unit of analysis—the focus point—as the social group, not the individual. From a social group perspective, roles link people, in particular through expectations about and understandings of roles. These understandings and expectations are the main motivators of and explanations for behaviors.

To understand this theoretical perspective better, think of just one of the roles that many readers may have: the role of student. (All of us have multiple roles, so, for a complete picture, we would need to analyze all the person's roles. For the moment, however, we focus on just a single role.) As a student, you understand that you have certain obligations and responsibilities, such as attending class, reading, asking questions, writing papers, and taking examinations. In return, others in related roles, in particular your instructor or professor, also have obligations and responsibilities related to your role, such as assigning readings, giving lectures, passing on information and ideas, reviewing your written work, correcting your examinations, and providing grades. These *role expectations* are very real and order the interactions of students and professors.

To understand just how real these expectations are, think of instances in which some of the role expectations are not being met. Imagine a case in which the professor never attends class, never lectures, and never answers questions. You, in your student role, are justifiably upset when this occurs because this social interaction is not a balanced one: The professor will still assign a grade. One way to resolve the issue would be to drop the class and thus end the role interconnection. (Note in this example that the disquietude the student experiences does not relate to anything "on the inside" of either the student or the professor, such as personality issues; instead, it is the result of mismatched role expectations.) Role expectations are also known as *norms*—socially defined standards of behavior that guide individual actions. In the example above about the student and professor, the sample obligations and responsibilities (for example, taking tests) are norms that relate to the acceptable performance of these roles. Without norms, we would not know what should properly occur between a student and a professor, and people in these roles would not know how they were expected to behave. Some roles have very clear norms on which everyone in a social group agrees (for example, the role of automobile driver), and others (for example, citizen of the world) have poorly defined norms about which different people may disagree.

In Chapter 9, we first discussed norms as one factor that can affect people's decisions to protect themselves from HIV infection. In the social interaction involving the roles of lovers, there are norms about behaviors that are part of these roles (such as having sexual encounters) and, for some people, norms about using protection (such as condoms) during these encounters. HIV/AIDS prevention programs, like those we described in Chapter 9, can attempt, as one of their aims, to make particular norms more conscious for people and also more definitive (that is, condoms must be used for vaginal and anal sexual intercourse).

Role theories will be useful to us later in this chapter when we consider such human issues as role conflicts for PWAs or between PWAs and others with whom they interact. These types of theories will also help to explain some societal issues, to be covered in Chapter 11.

Cognitive Theories

The other set of theoretical perspectives we will discuss is very different from the set of role theories. In contrast to role theories, cognitive theories focus on the conceptions inside people's minds, not factors (such as roles) in the outside world. The basic idea behind cognitive theories is that mental conceptions (also known as *cognitions*) give us a framework both for interpreting experiences and for shaping our actions.

Cognitions are mental representations of knowledge or thoughts, not feelings or actions. Related individual cognitions are combined in our minds into cognitive structures, known as *schemas*. *Schemas* serve as reference points for organizing past experiences or interpreting new experiences, or they serve as templates for activating new ideas. These mental representations undergo mental processing that eventually leads to action or inaction. Although our minds process schemas of all types and on all subjects, we are most concerned here with social cognitions, those that focus on people in our lives and our interactions with them.

An example helps to illustrate the perspective and utility of cognitive theories, with particular attention to social cognition. Sora and her friend Joy see a poster describing a panel discussion by PWAs that is to occur later in the week. Assume that Sora has taken a class on HIV/AIDS and therefore has some experiences and knowledge related to AIDS. Joy, however, has not taken an HIV/AIDS class and has only seen someone with HIV/AIDS on an occasional news report on television. What do Sora and Joy think when they see the poster about the PWA panel? Each of them unconsciously draws on the schemas in her mind to envision what the PWA panel might entail and, based on this, decides whether to take action. It is important to remember that the mental processes we describe next happen almost instantaneously. To understand where these processes come from and how they operate, however, we need to slow down the mental process. Consequently, in the following descriptions, the mental processing steps are set out in sequential fashion and with a self-consciousness that is not typical of their normal instantaneous operation.

Let us consider the cognitive processes for Sora first. Sora has studied HIV/AIDS, has seen videotapes in class about PWAs, and has met several PWAs. All these experiences have resulted in cognitions for her; these cognitions have been stored in her mind and united into a schema about PWAs. When Sora sees the poster about the panel, this schema is activated, along with other related schemas (for example, her schema about a panel). These schemas make Sora think of a small group of people who reflect the HIV situation as she understands it in her community: somewhat more men than women, somewhat younger than a random group of people, perhaps one drug user, and probably several people of minority racial/ethnic status. Otherwise, Sora pictures a panel that looks like almost any other panel. She also imagines that the panel members will behave like most other panels, with the notable exceptions that they will be more personally revealing and candid and that they will discuss matters that are not usually part of most panels (e.g., sexual behavior, family reactions, finances).

Joy has a much different picture in her mind when she reads the poster. Because she has had no direct education about HIV/AIDS and no direct experience with anyone who is HIV positive, she does not have the well-developed schemas on which Sora can draw to interpret the poster. Instead, Joy draws on (almost unconsciously and instantaneously, remember) schemas about a panel and about HIV/AIDS based on a few television images. Thus, Joy pictures a group of very sick people, some coughing, some in wheelchairs, one very angry and belligerent.

Who do you think is more likely to attend the PWA panel? Sora is probably your choice. Why? She has a clearer idea about who and what the panel will involve, and these cognitions have positive or neutral connotations. For Joy, the cognitions are vague, and they have negative connotations (sickness, anger, and bitterness). Unless other positive cognitions are activated for Joy (perhaps the opportunity to have dinner before the panel with her friend Sora), it is unlikely that Joy will attend the panel.

Note how, in this example, cognitions and schemas are brought instantly into play in interpreting new reality. Then, from this interpretation, the likelihood of experiencing the event begins to vary. Joy would probably benefit more from attending the PWA panel than Sora because the experience would be a very new and different one for her. She likely would be struck, for instance, by the "normal" appearance of the panel members. But, because of the way cognitive processes work, Joy is unlikely to attend. Even if her friend Sora tries to convince her to attend, the images in Joy's mind are difficult barriers to overcome. Remember, too, that these images may be just outside of consciousness to Joy, and to the extent she is aware of them, she may be reluctant to share them with Sora. It is easier for her to tell Sora, "I'd love to go, but I'm busy that evening."

Before we leave cognitive theories, there are a few additional concepts to discuss. First, social cognitions are unique in that the objects on which they are based—people—have a tendency to change and add a new external reality with which our minds need to deal. If our friend Duane, for example, changes jobs or changes wives or even changes haircuts, our mentally stored schema related to him will be somewhat at odds with the new reality he presents. The stored schema will need some adjustments, but these adjustments are made against the background of the old schema, which exerts a conservative status quo pressure. Every time a schema is activated, it undergoes some adjustments and expansion and is then restored in a strengthened form. Frequently activated schemas, such as those related to our own self-concept, are well established in our cognitive structures and become better established—and, therefore, more difficult to change—with each new activation.

Second, our cognitions—and especially our social cognitions—form a generally balanced and ordered whole. Think again about the example of changes in our friend Duane. Our schema of Duane may have both positive and negative components, but these cognitive components are balanced to create a schema that has an internal consistency. Imagine that we learn from a friend that Duane has been arrested for bank robbery. This news, completely contradictory to any of our schema components, presents a *cognitive dissonance*—an imbalance in our cognitive conceptions—that needs to be resolved. Our first reaction might be that Duane was mistaken for someone else, which would resolve the cognitive dissonance. Another reaction could be that Duane was at the wrong place at the wrong time: He was at the bank to make a deposit but was kidnapped under pressure and forced by others to be the front man in the robbery. Either of these explanations brings our old schema of Duane (as a law-abiding person) into line with the new reality and puts the schema back into balance. If neither of these explanations proves to be true, however, and Duane is convicted of bank robbery, we must deal with a new reality, and our mental schema of Duane will likely change toward a more negative one. If the negative aspects become too great and there is too much cognitive dissonance, there will be strong pressure inside of us to take some action to end the dissonance, perhaps by ending our relationship with Duane. The important aspect to note in this example is that unbalanced schemas have the potential to cause changes of various sorts.

Cognitive theories will be useful to us later in this chapter when we consider such human issues as self-esteem and self-concept for PWAs or those interacting with them. The cognitive perspective will also help us understand the societal issues covered in Chapter 11. We had a preview of some of these issues in Chapter 9, in our discussion of the difficulties of getting people to change well-established attitudes, beliefs, and behaviors that put them at risk for HIV infection.

With these two theoretical perspectives as background, we now are ready to explore and better understand some of the human and societal dimensions of living with AIDS. The remainder of this chapter is devoted to human dimensions and in particular focuses on people living with HIV/AIDS.

Human Dimensions of HIV/AIDS

Those living most directly with HIV/AIDS are people infected with the virus. These individuals have significant challenges—psychological, social, and physical—in dealing with HIV. They also have significant opportunities not open to those who are HIV negative. In this section, we explore these challenges and opportunities, using perspectives from both role theory and cognitive theory.

Confronting the News of Infection

Many HIV-positive people first learn about their seropositive status when they have an HIV test. Others receive this news when they develop their first opportunistic infection and, in the course of working with their doctor to diagnose this physiological change, learn that they are HIV positive. In both cases, although the mental processes can vary, dealing with the new information concerning HIV infection can be a significant challenge. In each case, however, the news of HIV infection puts major pressure on the individual's self-concept.

Self-Concept

The self is probably one of the most developed schemas each of us has in our mind. Although we may not often be aware of the fact, our self is a social construct, developed and maintained in interaction with others. If those around us make sudden changes in their behavior toward us, we quickly become conscious of the importance of others' appraisals in how we view ourselves. Usually, however, changes in others' behaviors toward us are subtle and do not occur all at once. Consequently, only minor adjustments are usually made in our self-concept.

As we discussed in Cognitive Theories, schemas are reinforced each time they are activated. The self-schema, because it is regularly activated, is one of the strongest schemas in most people's cognitive structure. Parts of the self-schema can vary, and minor adjustments are regularly made, but by and large we have a constant and firm *self-concept*. One part of the self-schema that varies is our *self-esteem*, our positive or negative evaluation of ourselves. A series of good experiences can boost our self-esteem, and a series of bad ones can dampen our self-esteem. Nonetheless, most people have a fairly constant appraisal of themselves, built up through experiences that anchor their self-esteem at a high or low level.

Consider, then, the cognitive situation Fernando confronts when he learns he is HIV positive. This news is a direct and significant threat to his self-concept. Making matters more difficult is the fact that Fernando does not seem any different than he was before he learned the test results: He is still healthy and confident, a good swimmer, a bad singer, a good son, a thoughtful friend, and so forth. But, out of the blue, he receives news that potentially puts all of this in jeopardy. It is understandable, from a cognitive perspective, that Fernando's first reaction is denial. This new piece of information does not fit into his self-concept. Indeed, to accommodate the reality of HIV-positive status and bring his self-concept back into balance, Fernando will have to change many aspects of his self-concept—something that none of us does quickly or easily. Moreover, other than the HIV test counselor, who does not play a large role in Fernando's life, no one else knows this information at the outset. Fernando may simply decide that the test result is wrong, thereby denying the new information and preserving his self-concept. Good HIV test counselors are prepared for this reaction and realize the tremendous pressure that exists in all of us to maintain our current self-concept. The HIV test counselor may encourage Fernando to have a second HIV test, not because she believes the test result is wrong (although there is a very very small chance of this; see Chapter 8, p. 148) but because it will give Fernando time to think about the new information and add another piece of information that supports the first.

One reaction the counselor wants to avoid is forcing Fernando to fit this new information into his self-concept. It may cause him to turn against the news in understandable but dangerous ways. For instance, Fernando could set out to reinforce his current self-concept (that he is HIV negative) by engaging in risky sexual behaviors, which, to him, "proves" he is not HIV positive and, at the same time, reinforces his concept of himself as a sexually desirable man. Other potentially troubling cognitions (for example, that he is socially and morally irresponsible to put others at risk for HIV) may, for the time being, be conveniently avoided, because, if Fernando truly believes he is HIV negative, he sees no issues of social or moral irresponsibility. From Fernando's perspective, you can begin to see how cognitively complicated the situation quickly becomes and how certain decisions and actions—sometimes ones that seem to others to be wrong, irrational, or immoral—can lessen the complexity and mental pressure that Fernando experiences.

Although many people who learn they are HIV positive (without any other outward signs of AIDS) may go through a period of denial, most do not act out their anger at the news in this way. Instead, they do what most of us do when a very serious threat to our self-concept arises: They talk to their closest, most important friends.

Because we define ourselves through our interactions with others, we also redefine ourselves this way. This is not an easy step, however, because we are not sure how our friends will react. Will they believe we are diseased and never see us again? Will they still love us, even though they do not fully understand HIV/AIDS? Will they still have sexual relations with us if we use protections (such as condoms)? These questions raise issues of role experiences and expectations, which can be understood more completely if we shift to the role theory perspective.

Role Experiences and Expectations

As discussed earlier, we interact with others in terms of roles that are more or less well defined. Among people with whom we interact regularly, these roles are well defined, although they usually are not explicitly articulated. Instead, the dimensions of the roles are created in interaction over a long period of time, and adjustments are made as the individuals change and grow. Sometimes, these adjustments make the roles more formal and distance the individuals; at other times, the adjustments open up new dimensions to the roles and increase the closeness between the people.

An individual who is dealing with the news of HIV infection commonly turns to good friends and family. Just as this news is threatening to his or her self-concept, the new information about HIV infection can be threatening to the role expectations that good friends and family share. Frequently, this one piece of news raises other issues that also may be new. The more issues that are raised, the more potentially threatening the news can be to the role experiences and expectations of a well-established relationship.

Consider the case of Trina. She learns that she is HIV positive and decides to share this information with her best friend, Suzi. Suzi is likely to ask how Trina contracted HIV, and this question triggers discussion of the topics of Larry and of the sexual relationship Trina and Larry were having, a fact unknown to Suzi. What Suzi learns may not fit with her current expectations about the roles she and Trina have had together. However, Trina's disclosure of this news to her good friend reinforces the importance of and the emotional support from the relationship. Fortunately, at sensitive moments like this between two good friends, each draws unconsciously on her well-developed understanding of the role expectations and behaves in keeping with these expectations. It is rare, for example, for Suzi to announce to Trina, following Trina's disclosure of her HIV status and her sexual relationship with Larry, that she is ending the relationship with Trina. More likely, she will show great support and probably great emotion, in part because she too is confused by this sudden news and about how to incorporate it into her expectations about Trina and about their future relationship. From Trina's viewpoint, the support and emotion are much-needed reinforcements that reaffirm the relationship and the roles. Both understand that the future holds unknowns but also that they will work through the new realities together, drawing on the reservoir of support, respect, and love.

Although the disclosure conversation with good friends generally proceeds in this manner, this is not always the case with more casual friends, work colleagues, and sometimes family. Let us consider the case of casual friends and work colleagues first; families require special attention. The role experiences and expectations we develop with casual friends or work colleagues usually are much more limited and less complex than the roles we develop with good close friends. We may, for example, work very well with a particular colleague, precisely because we have carefully circumscribed our role interrelationships. Indeed, we can have good work relationships with people with whom we disagree on important social or political issues by limiting the dimensions of our roles with them.

It is these very limitations, however, that present the challenge for an HIV-positive individual regarding disclosure of HIV status. The HIV-positive individual is not sure whether the role expectations can incorporate the news of HIV or even whether the role expectations will preclude an immediate, negative reaction, in contrast to the superficially pleasant demeanor that is part of the roles most of us have with casual friends and colleagues. However, there is the great potential for a significant expansion and deepening of the role expectations with particular individuals, such as a work colleague who now discloses that her husband is HIV positive or a casual friend who discloses that he has been HIV positive for several years.

The problem for the HIV-positive individual is that he or she cannot know in advance how the limited role expectations with casual friends or colleagues will accommodate the news of HIV-positive status. In all likelihood, the HIV-positive individual will take cautious steps toward greater disclosure, wanting to avoid rejection or negative reactions that would only add to the pressure on his or her self-esteem.

We said that families need special consideration. We are born into family roles of son or daughter, brother or sister, which are laden with expectations and defined through intense experiences as we grow up. The extent of the son/daughter and brother/sister role expectations varies for each of us, and the centrality of these expectations to our own self-concept also varies greatly. These roles are very important to some of us, with interrelationships similar to those we described for close friends. Others of us are less involved in these particular roles, with interrelationships more like those with work colleagues. Because of these variations, some newly identified HIV-positive individuals will consult their families first, and others will avoid telling their families this news until absolutely necessary—and perhaps not even then.

Accepting the Reality of Infection

Each HIV-positive individual eventually begins to accept the reality of HIV infection. In different ways for each individual, the self-concept is adjusted and roles are changed. The process of acceptance involves a reshaping of the individual's self-concept and self-schema and a reassessment of role expectations and responsibilities.

This process of cognitive and role adjustments involves both risks and opportunities. The risks include dealing with issues such as unresolved tensions in relationships, unrealistic hopes and dreams, unattainable goals, and unpleasant personality dimensions. There are opportunities as well, some very significant. As HIV-positive individuals reshape their self-concept, they are able to focus on their strengths and anchor their goals in ways that are most satisfying to them. No longer, for example, must they strive to achieve goals that became incorporated into their self-concept through pressure from family and others but were not truly their own. HIV-positive individuals also have the opportunity to leave unsatisfying roles and adopt new ones. Some of these new roles might include public presentations to groups about HIV; work with people with AIDS in their terminal phases; volunteer work on causes that are important to them; or more individual but equally satisfying roles such as amateur artist, gardener, or daily exercise walker.

The process of adjustment is usually not easy and, for many HIV-positive individuals, is one that is never completely resolved. The irregular and unpredictable pattern of positive and negative physiological changes that occur over the course of seropositivity can give the HIV-positive individual renewed hope one day and dejection the next. On the longer term, new drug therapies have changed HIV from an acute disease that likely ends in death to a chronic condition that can be managed for a very long time. This reality adds both hope and uncertainty, but it is a reality that is only possible for those with the financial resources to pay for the drugs and care.

The path of acceptance is different for each individual, depending on a number of factors, such as changing physiological realities, personality tendencies, shifting financial matters, and social support. As an illustration of the effect of one of these factors, consider Scott, a man who is very angry at having a life-threatening condition. Sometimes this anger is directed at particular people whom Scott blames for causing him to have AIDS or at people with whom Scott has old grudges and complaints unrelated to his illness. Other times, Scott's anger is more diffuse and nonspecific; it can be focused at the world in general or at those who happen to be around him. In this latter case, Scott unintentionally hurts those closest to him and inadvertently threatens their continued support and understanding. Those around Scott need to realize that they are being scapegoated and not personally targeted.

Although people with other life-threatening diseases (for example, cancer) also have a difficult path of acceptance, those with HIV/AIDS frequently have added challenges due to the *stigma* associated with AIDS. To understand this difference, consider those diagnosed with breast cancer, for example. These people are not stigmatized (that is, judged negatively) solely on the basis of having cancer. They do not lose their jobs or their lodging because they have a cancer diagnosis. Those who are HIV positive or diagnosed with AIDS are, too often, treated differently: They are judged and even censured by some people for having the disease; they might be fired from their jobs; or they might lose their lodging. There have been, and unfortunately continue to be, instances wherein those with HIV have had their houses set on fire or have been physically attacked because of their HIV status. Although antidiscrimination laws and policies have been changed in some places, mainly bigger cities in more developed countries, in many other places, those with HIV/AIDS are still stigmatized when they reveal their condition.

In confronting and accepting the reality of HIV, the HIV-positive person now has a very useful resource that did not exist in the early 1980s and 1990s: Internet-based support and information groups. Not only do HIV-positive people use the Internet to follow developments in HIV research, therapies, and care, they now also can access online support groups of people just like themselves who are dealing with the realities of HIV. The Internet allows people to be silent, anonymous observers, learning more and observing how others are dealing with HIV, or to be active participants, engaging in discussions and exchanges about the particular issues that are most relevant to themselves. Although there are websites focused generally on HIV issues for all those with HIV, the ones focused on particular subgroups of HIV-positive people are more useful in terms of dealing with the more personal issues related to self-concept and self-esteem. There are, for example, websites related to individual characteristics of HIV-positive people, such as women or different racial/ethnic groups, or those related to interests or special needs of HIV-positive people, such as women with children or the care available to those in particular geographical areas. See the book website related to this chapter for examples of sites focused on youth and women, as well as another for Asian-speaking PWAs (in several different languages).

The constant possibility of rejection, by those in general society, of individuals who are HIV positive or have AIDS affects the progress of personal acceptance of HIV and the actions that HIV-positive individuals take. This challenge can be illustrated by two examples, one with an individual focus and one with a group focus.

Greg Louganis: An Individual Dilemma

U.S. Olympian Greg Louganis was the winner of double gold medals in diving at both the 1984 and 1988 Olympic Games. In 1988, at the start of the Olympic trials, Louganis knew that he was HIV positive.[2] He had accepted this reality and moved beyond it. He focused all his energy and attention on the 1988 competition, knowing that he was not at his peak physiological performance level (due to aging, not to HIV status). Louganis was comfortable enough with his HIV status to share his diagnosis with his coach but not with others, due in large part to the strong stigma associated with HIV/AIDS in the late 1980s. In addition, he believed his HIV status was irrelevant to the conditions of a diving competition (the Olympic Committee required no disclosure of HIV status for any athletes at that time and still does not), and he believed that he would not put anyone else involved in the diving competition at risk.

During the diving competition, Louganis unexpectedly hit his head on the diving board, cut his scalp, and bled into the pool as he tumbled into the water (Figure 10-1). Louganis had never considered this possibility but immediately realized the risk his bleeding cut could present to the team doctor who was about to put temporary stitches in his scalp so that he could continue the competition. He also wondered about the

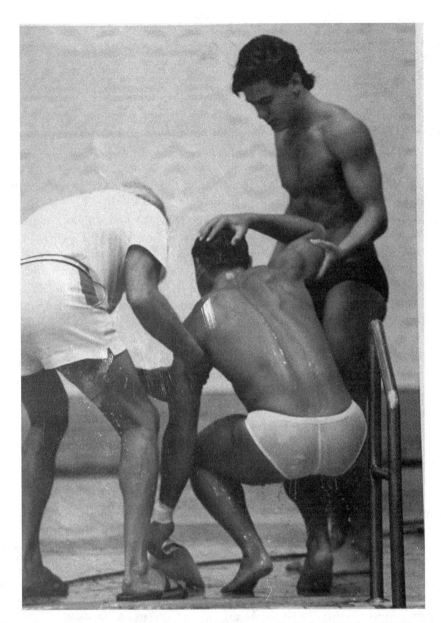

Figure 10-1   Greg Louganis after hitting his head on the diving board in the 1988 Olympics diving competition. (© Bruno Torres/AP Photo.)

risk to other divers from the blood in the pool. (Louganis did not need to worry about the latter issue; the chlorine in a swimming pool quickly kills HIV, which would have been greatly diluted in any case by the water.)

From our earlier discussions of self-concept and self-esteem, we can see how Louganis faced strong and complex psychological pressures in deciding whether to tell the doctor his HIV status. On the one hand, he could reveal his HIV status and refuse to let the doctor sew his scalp. This would throw the diving competition into turmoil (and, to some extent, the entire Olympic Games) and end the possibility that he would win a record-breaking second set of gold medals (he had 20 minutes to be back on the board for his next dive). Like those of other Olympic competitors, Louganis's self-concept and self-esteem were overwhelmingly tied to his competitive ability and success. Forfeiting his opportunity to set another Olympic record would be a great blow to his self-concept. In addition, disclosing his HIV status in such a public manner to people around the world and the censure he probably would have to endure would further undermine his self-esteem. On the other hand, Louganis could say nothing and not reveal his HIV status to the doctor. This would put the doctor at potential risk for HIV. With all these conflicting and complex psychological pressures and in a dazed physiological state from the blow to his head, Louganis later said he was "paralyzed with fear" and so he said nothing.

The team doctor was going through his own risk assessment and decision making (see Chapter 8). He thought about using plastic gloves for protection, but he knew none were available and that he had only 10 minutes to complete the stitching after he finished his examination of Louganis. He made the decision to proceed with caution. (The doctor correctly knew that touching HIV-infected blood alone is not a sufficient condition for HIV transmission; he would have had to have opened up a path of transmission by cutting or pricking himself; see Chapter 7.) The doctor subsequently tested HIV negative.

Louganis went on to win double gold medals, but he also had to deal with new uneasiness about the realities of HIV infection. Following the 1988 Olympics, Louganis kept this uneasiness to himself and a very small group of friends. By releasing his autobiography, in which he acknowledged publicly not only that he is HIV positive and has AIDS but also that he is homosexual, Louganis moved to another deeper level of acceptance of the reality of infection, as well as of his sexuality. In publicly sharing this information, Louganis sent a message to others about the need to stop HIV infection and the personal agonies caused by the stigma still associated with HIV/AIDS.

ACT UP: A Collective Response

The acceptance of the reality of infection takes many forms for those with HIV. In the earlier section, Accepting the Reality of Infection, we described Scott, whose anger about the disease affected his perceptions and the reality of those around him. In some HIV-positive individuals, this anger is fueled by the explicit and implicit rejection people experience because of the stigma associated with HIV/AIDS or by prejudices associated with related issues, such as homosexuality or drug use, that sometimes come to the fore when HIV occurs. Because the rejection arises from prejudice and discrimination (subjects to which we turn in Chapter 11), the HIV-positive individual is understandably frustrated and angered by actions that are unfair and unjust. The resulting feelings of anger and injustice in some HIV-positive individuals provide powerful fuel for actions of different types and, in some cases, with unusual effectiveness.

These transformations of anger into action provide one example of the kinds of unique opportunities PWAs have to effect change. Consider, for instance, the emergence of the ACT UP organizations in the United States and later abroad. *ACT UP* originally stood for "AIDS Coalition to Unleash Power." Now, it is more generally associated with groups of PWAs or their supporters who organize and implement public demonstrations to increase awareness of AIDS issues. Their tactics are usually tough: chaining themselves to government buildings, disrupting church services, throwing vials of bloodlike substances, and chalking outlines of dead people on public sidewalks. These actions usually involve a public display of anger and a confounding, if not an outright rejection, of role expectations. As we have discussed, sudden one-sided changes in role activities and expectations are uncomfortable for those involved in social interaction. These changes, however, get people's attention, and ACT UP has not missed this opportunity to raise people's consciousness about AIDS by jarring and challenging role expectations and experiences.

Consider one famous ACT UP church demonstration that occurred in New York City at a Catholic cathedral. Demonstrations during church services (including such activities as shouting, passing out information, and lying on the floor) are not part of most people's role expectations of how those attending Catholic church services should behave. Imagine the attention the small group of ACT UP demonstrators received—as well as the anger they generated—when they began their demonstration. The media covered the event, which was broadcast around the world; a documentary film was even made about the planning and execution of the demonstration.

At the individual ACT UP demonstrator level, this event provided a vehicle for release of personal anger and for action in the service of greater AIDS awareness. For the individual non–ACT UP churchgoer, the event provoked reactions from irritation to frustration and anger. At the larger community level, the demonstrators accomplished at least one of their goals—greater attention to AIDS in New York, in the United States, and around the world—but whether they gained more than they lost in public support was unclear. Even among the New York City ACT UP group, the decision to hold the church demonstration was controversial and not endorsed by everyone.

Other ACT UP–motivated activities have had clearly positive community outcomes. Members of ACT UP and other AIDS activist organizations pressured the National Institutes of Health (NIH) in Washington DC to speed up the official approval process for potentially life-saving drugs and therapies. As they pointed out, in polite meetings with NIH officials and in not-so-polite demonstrations in front of NIH headquarters, they were dying, and they could not wait for the 5- to 10-year approval process. The AIDS activists wanted to be allowed to make the decision themselves about whether to participate in experimental therapies. NIH expanded its procedures for approval of new drugs and in some instances now allows greater availability of potentially beneficial drugs through local physicians in community-based trials. This ACT UP action has benefited those with other diseases too; the U.S. Food and Drug Administration has accelerated its approval process for drugs of all types.

These examples demonstrate some of the larger scale opportunities open to HIV-positive individuals as they progress in dealing with the realities of HIV infection. Many smaller scale opportunities present themselves as well, including speaking on panels about HIV/AIDS or volunteering with community groups on activities related to HIV/AIDS or on unrelated activities, such as education or environmental protection. In addition to the benefits to others from these large- and small-scale actions, the HIV-positive individual gains increased self-esteem from achievements in new roles.

Opportunities and Challenges of Drug Therapies

Various drug therapies have been developed for those with HIV/AIDS, adding hope but also complications to the challenge of living with AIDS. These therapies and treatments typically are introduced with great expectations and usually raise hopes in those with HIV/AIDS, only to result in some disappointment later, as the limitations of the drugs become apparent. AZT is an example of one of the earliest drugs (see Chapter 5). One of the newer developments on the drug therapy front has been the success of the "triple combination therapies" in reducing viral loads to undetectable levels (see Chapter 5). At the outset, scientists cautioned that we needed more and longer-term data to be sure of this effect. Those in the AIDS treatment community and those with AIDS, however, reacted quickly and very positively when the first scientific results were announced.

This reaction is understandable for all the reasons we have been discussing related to self-concept and self-esteem. One initial effect of the combination therapies is very clear: Many who take the drugs feel much better physically if they have previously been experiencing AIDS symptoms. Some report never having felt better and returning to their regular degree of health before HIV infection. It is easy to imagine the psychological effect these physiological changes would have: new feelings of control, increased hopefulness, thoughts of living far into the future, the renewed ability to set goals and to dream of new possibilities. The absence of all the HIV- and AIDS-related physiological cues reenergizes the mental schemas held before the onset of HIV/AIDS.

Alas, the situation is not so simple. As with most drug regimens, challenges are associated with the combination therapies. Some people experience constant unpleasant side effects from the drugs, including nausea. Also, not following the prescribed regimen can greatly reduce the drugs' effectiveness.

The challenge for the largest number of people with HIV/AIDS, however, is obtaining the drugs in the first place. Combination therapies are expensive: Estimates of the yearly cost are about $15,000. Those who have health insurance during their AIDS symptoms period (a minority of those currently infected) are protected from these large costs. For most, however, including those without insurance, without personal resources, and on Medicare, the cost is high, often prohibitively so. In the United States, we know that the AIDS epidemic has moved into socioeconomically disadvantaged populations that do not and will not have access, either directly or through public or private social services, to the funds for combination therapies. In developing countries, where the epidemic is spreading most rapidly and where money for care is limited, few resources are available for combination therapies.

In this chapter, we presented two important frameworks to assist in understanding the human and societal dimensions of living with HIV/AIDS: role theories and cognitive theories from social psychology. Then, using concepts from these theories, we explored some of the human dimensions, particularly those focused on people with HIV/AIDS. In the next chapter, we expand our perspective to look at societal dimensions of HIV/AIDS.

Notes

1. The discussion here is based on K. Deaux, F. C. Dane, and L. S. Wrightsman, *Social Psychology in the '90s*, 6th ed. Pacific Grove, CA: Brooks/Cole; 1993. The reader is encouraged to consult this book for fuller discussions of the social psychological theories and concepts presented in this chapter.

2. See Greg Louganis's autobiography for more details: *Breaking the Surface.* New York: Random House; 1995.

---

http://biology.jbpub.com/fan/aids/6e/

Connect to this book's website: http://biology.jbpub.com/fan/aids/6e/. The site features summaries of the main points from each chapter, links to important AIDS-related websites, and short-answer-style review questions for each chapter.

CHAPTER 11
Living with AIDS: Societal Dimensions

Theoretical Concepts
Prejudice
Discrimination

Societal Dimensions of HIV/AIDS
Needle Exchange for Injection Drug Users
HIV Prevention for Teens
Healthcare Practices

Individuals, communities, and societies are intertwined and interrelated; none of us exists in a vacuum. Because people are parts of communities, HIV infects not only individuals but also communities and societies—which are, in turn, affected by these social groupings. In addition, the commonly held attitudes and norms in a community affect how those with HIV/AIDS are treated and how HIV is—or is not—prevented. In this chapter, we explore some of the societal dimensions of living with AIDS/HIV.

At the level of nations, societal norms are critical in recognizing and acting on HIV/AIDS. This was demonstrated dramatically in the early to mid-1990s when certain nations maintained that HIV/AIDS could not, and therefore did not, exist within their countries. Because AIDS did not exist, they saw no need to take action to prevent it. Unfortunately, this attitude fostered the spread of HIV/AIDS in these countries. At this point, every nation recognizes the reality of HIV/AIDS, but some are still resistant to accepting the accurate reasons for the presence and spread of HIV infection, to treating those with HIV without discrimination, and to using proven therapies and regimens for those living with HIV/AIDS.

In Chapter 10, we focused on PWAs—people living directly with HIV/AIDS—and the challenges they face personally in dealing with AIDS. Societal issues, as we saw, cause some of these challenges. For example, society's judgmental attitudes about those with AIDS make it more difficult for a PWA to accept the reality of infection. In this chapter we expand the focus beyond PWAs to everyone else who is living with HIV/

AIDS, although often indirectly as part of the larger society. We concentrate here on two general issues—prejudice and discrimination—that underlie many of society's specific actions related to HIV/AIDS, from healthcare practices and policies to laws related to HIV/AIDS. We illustrate the general principles with three examples of these specific actions: needle exchange for injection drug users, HIV prevention for teens, and healthcare practices.

Theoretical Concepts

Two concepts are important to understand before we look at societal issues and AIDS. As we did in the last chapter, we have drawn on the field of social psychology to provide the foundation for our understanding. In particular, we focus on two concepts that deal with people's preconceived attitudes about others and how these affect their behaviors. First, we look at the concept of prejudice; we then turn to its related partner, discrimination.

Prejudice

Prejudice is a biased attitude toward a group of people. It arises from schemas (see Chapter 10, p. 174) in people's minds that are associated with particular groups of individuals. Prejudices can be either positive or negative. Think about some positive prejudices you may have. To do this, answer this question: What groups of people do you tend automatically to believe as good? One person might have a positive prejudice toward healthcare workers; another may have a positive prejudice toward Asians. Now, think about some negative prejudices you have. What groups of people do you tend to think of negatively? The same groups toward which one person has a positive prejudice (healthcare workers or Asians, in our earlier example) can be the focus of negative prejudices for another person. This illustrates that prejudices are less about realities anchored within a group and more about what is in the mind of the person making the judgment. The person making the judgment is thinking first about "groups" of people, not about particular individuals. In addition, the person's perception of this group becomes the reality. Racial or ethnic groups are common targets of prejudice—frequently negative—because their members have easily identifiable common characteristics, which makes it easier to group people, and because there has been a history of friction between different groups, which provides some "data"—albeit selective—to frame the perception.

Stereotypes form the cognitive basis for prejudices. Stereotypes are group-based schemas, as contrasted with individual-based schemas (such as, for example, your schema about your mother). Stereotypes can be either positive or negative, but like our other social schemas, they usually reflect society's general appraisal of the object of the stereotypes. For example, our stereotype of newborn babies is quite positive: cute, content, cuddly objects. (If you doubt the power of this stereotype, try telling a new mother that her baby is not cute, is ill-tempered, and repels you.) At the other extreme, our stereotype of vagrants is negative: shiftless, ill-kempt, bad tempered, dangerous. (If you want to test the power of this stereotype, try reacting in a very obviously positive manner to vagrants on the street in the presence of "regular" community members; you likely will be the butt of negative comments from these passersby.)

Because they are group based, stereotypes are usually not accurate in their details when applied to individual members of the group. Not all newborn babies are cute, and not all vagrants are bad tempered. When we have opportunities to interact with particular individuals (our brother's new baby, for example, or the vagrant we always pass in the park), we can develop individually based schemas that may begin with the stereotype but are changed to relate to the specific individual. The original stereotypes are usually not changed to accommodate the new individual-specific experience unless we have many experiences with many new babies or many vagrants. You may have heard people say something like "Sam is a nice guy who is down on his luck; he's the exception that proves the [vagrant] rule." This explanation resolves the apparent contradiction between the positive appraisal of Sam and the negative appraisal associated with the vagrant stereotype; it also reinforces the negative evaluation of vagrants.

With this general background, let us consider the relationship of prejudice to HIV/AIDS. Our main focus in regard to societal impacts of HIV/AIDS is the effects of negative stereotypes and negative prejudices. Think back to 1981. Although not yet labeled, HIV/AIDS first appeared in reports about unusual life-threatening diseases in homosexual men (see Chapter 1). As the epidemic grew in the next few years, more and more male homosexuals died, and the general population came to learn of HIV/AIDS.

How did the community react to the news of these strange deaths? As we learned in Chapter 10, when new information arises, people draw on relevant cognitive schemas to interpret it. Schemas related to death and dying prematurely were probably activated for most people, causing some fear in most of them. In addition, other related schemas were probably activated, such as those related to homosexual men. For most people, these latter schemas were based on no direct contact with homosexual men but instead on stereotypes, which in the early 1980s at the outset of the AIDS epidemic were generally negative in our society.

Prejudice against homosexuals initially fueled the reaction of the general populace to HIV/AIDS. For many people, a negative outcome (death) was occurring to a negatively evaluated group (homosexuals); consequently, it was easy to blame the victims of HIV/AIDS for their deaths. As the epidemic spread into the injection drug-using community, prejudice against this particular community by the general population further fueled the negative reaction to HIV/AIDS. By blaming these victims, people in the general population could explain HIV/AIDS as originating from these "bad" people. This relieved them of any fear that they would contract HIV/AIDS (because they were not members of these negatively stereotyped groups) and also relieved them of any responsibility for action toward these victims (because the victims had brought this on themselves).

As HIV/AIDS began to affect positively stereotyped groups, the situation became more cognitively confusing, and it was more difficult for people to reconcile the positive and negative aspects of their schemas. Babies were born with HIV, and hemophiliac children developed AIDS. Because of the positive stereotypes associated with these groups, these PWAs were not blamed but instead were called "innocent victims"; it was not their fault they had HIV/AIDS. Then, favorite movie and TV stars began to die from AIDS, and esteemed sports figures announced they were HIV positive. Now, women and minority group members are the fastest growing segments of the AIDS population. All these groups can have stereotypes associated with them, so parts of these stereotypes become mixed with the HIV/AIDS schema in our minds.

As HIV/AIDS has spread throughout society, it has become more difficult for most people to use negative stereotypes of certain groups to explain the epidemic, to distance themselves from it, and to justify inaction. Although the general level of prejudice against those with HIV/AIDS has lessened, there is still a negative stereotype associated with the disease. In some places, particularly larger cities where HIV/AIDS has been a major issue, people's HIV/AIDS stereotypes are more varied, with more positive and neutral views. In these cases, the more positive attitude regarding those with HIV/AIDS has allowed some public displays of support. AIDS walks occur in these places to raise funds and show support for HIV/AIDS programs of all types, from prevention to treatment and care (Figure 11-1). These walks are usually half-day events that feature individuals or teams of individuals, most of whom have solicited financial pledges from others, who walk along a designated course for several hours. In addition to raising funds and spirits among those who participate, the events provide a public display of positive attitudes and action toward HIV/AIDS that helps to lessen the negative stereotypes in the general public.

Discrimination

Prejudice relates to attitudes; discrimination relates to actions. *Discrimination* is any behavior toward an individual based only on the individual's membership in a particular group. In societal terms, discrimination is usually thought of as only negative actions, and we use this connotation. However, be aware that positive discrimination— favoritism shown to an individual based solely on group membership—is also possible.

Discriminatory behaviors can be directed at a particular individual or at groups of individuals, but in both cases it is the group membership that anchors the action.

Figure 11-1   The Boston AIDS walk, an annual 10-kilometer walk to raise money for AIDS research and services. (Courtesy of Pia Schachter/AIDS Action Committee of Massachusetts.)

To illustrate this, consider these examples. Joyce loses her job when her boss learns that she is HIV positive. Ramon's apartment lease is suddenly terminated after his neighbor tells the landlord that Ramon has AIDS. The legislature of a U.S. state passes a law that restricts HIV-positive individuals from certain social service benefits. In the cases of Joyce and Ramon, the action occurs solely because of each one's membership in the group of people who are HIV positive. In the case of the new law, the restrictions occur solely because of HIV-positive status and, when implemented by actions, affect individuals who are members of this group.

When a group of individuals is easily identifiable, isolated, and highly stigmatized, like the vagrants in the park we discussed earlier, prejudice and discrimination are usually open and clear. So many people in society are so certain that their attitudes are correct they do not see a problem in expressing opinions about and taking actions against the group or individuals within it. In many other cases, however, prejudice and discrimination are much subtler and people's true motives are covert. This is especially true when a number of conflicting schemas are interrelated. We learn, for example, that we are not supposed to judge groups of individuals by their racial or ethnic backgrounds and that it is not good to express prejudices about these groups. Nonetheless, prejudice and discrimination against racial groups occurs, although rarely with a clear link.

If the link is not apparent, how do we know there has been discrimination? The analysis of actions across a group of individuals is required to document discrimination. For example, among a set of otherwise similar employees (comparable performance evaluations, similar age) at a company, the termination of more HIV-positive employees compared with the rest of the employees in the set suggests discrimination. Although proof of discrimination in individual cases may be hard to document, the result—that more HIV-positive people are terminated—provides convincing evidence that HIV status is playing a role in the terminations. This example illustrates another truth: Proving discrimination that is covert can be a long, laborious process.

Societal Dimensions of HIV/AIDS

In the early years of HIV/AIDS, the most prominent societal aspects were negative ones, such as the effects of prejudice and discrimination in the treatment of those with AIDS and in the slow and ambivalent response to AIDS prevention efforts. At this point, 25 years into the epidemic, the societal dimensions are more complex, with a number of lingering negative aspects but with some positive elements as well. Earlier, we described the AIDS walks that are now common in major metropolitan areas. In addition to their fundraising benefits, these public displays of support show general society that there is a constituency who supports activities related to HIV/AIDS. These walks, however, are half-day events, so the picture of HIV/AIDS support they present is constrained, and their utility in changing societal attitudes is limited. If we look at other, more regular HIV/AIDS related activities, we see a more complex and ambivalent societal picture, with both positive and negative forces pushing and pulling for influence. Next, we describe two cases to illustrate this tension: laws related to needle-exchange programs for injection drug users and HIV prevention for teens. We conclude with a third case from the healthcare area that presents a more uniformly positive picture of how societal influences can assist those living with HIV/AIDS.

Needle Exchange for Injection Drug Users

As discussed in Chapters 7 and 9, we know how to prevent the transmission of HIV infection among injection drug users: remove the possibility of transporting HIV-infected blood from one person to another. Drug users can accomplish this either by not sharing needles or by cleaning their needles before reuse. The challenges for many drug users are that clean needles are not available, that they do not have the resources to buy them even if they are available, or that they do not have the materials or time to clean needles before reuse. Those working in HIV prevention with drug users have recommended "harm reduction" strategies that include clean-needle-exchange programs, also called syringe-exchange programs. In these programs, drug users can exchange used needles for new ones at no cost and with no questions asked.

A large-scale evaluation of a set of needle-exchange programs, funded by the private American Foundation for AIDS Research, provided conclusive proof of the effectiveness of needle-exchange programs in significantly reducing rates of new HIV infection. These programs decreased HIV infection by 30% in injection drug users and also increased the likelihood that they would enter drug treatment programs. At the same time, these programs did not cause people who were not drug users to become users, which is an often-expressed concern of those against needle-exchange programs.

Unfortunately, those with negative attitudes toward HIV/AIDS, combined with negative attitudes toward drug use and users, are preventing the implementation of these successful programs. Only a few cities, with special exemptions granted by state officials, have instituted needle-exchange programs. Repeatedly, legislators vote down or elected officials veto new laws that would permit these programs. The legislators and officials most often say that people do not want laws that endorse drug use. In fact, the proposed laws carefully avoid any such endorsement and focus only on needle exchange as a means of stopping the transmission of HIV. Some proposed laws also have included provisions to increase resources for drug treatment and drug use cessation programs as a way to ameliorate the legislators' criticisms of and concerns about these programs.

The primary underlying problem is general prejudice against drug users, which results in discriminatory actions. Most people in society do not support using drugs, particularly injection drugs, and therefore do not support any actions that would benefit drug users. Some people are so prejudiced they are comfortable with the status quo: letting drug users share infected needles and increase their risk of HIV and of dying. For legislators, the societal attitudes surrounding this issue are complicated because the reason for the needle exchange relates to preventing the transmission of HIV/AIDS, a goal that nearly everyone in society favors. Even people who might still have negative attitudes toward those who have HIV/AIDS are in favor of stopping its transmission. So, legislators and public officials are in the difficult position of trying to balance conflicting societal attitudes: One group supports programs that can help stop the transmission of HIV/AIDS (namely clean-needle-exchange programs), and another group opposes these measures. The situation is even more complex because, at this point in the HIV/AIDS epidemic, negative attitudes toward those with HIV/AIDS are rarely publicly expressed, but these attitudes still operate under the surface for some in the general public, indirectly fueling the degree of resistance that is expressed.

How can this situation be changed? As we know from our discussion in Chapter 9, changing a person's attitudes and behaviors related to his or her own personal HIV risk is difficult; changing a person's attitudes and behaviors related to others' HIV risk is as difficult. In the case of drug users, we would need to educate the general public in much more detail about the reasons for and realities of drug use. It is unlikely that someone who is unsympathetic to drug users would pay much attention to our educational program. Alternatively, we could try to explain how ignoring the drug-using community only causes other problems (such as increased crime and violence) and results in costly actions (such as more police). Again, it will be difficult to convey this message to someone who has already made up his or her mind and has a strong stereotype to justify it. The most productive approach is the slow expansion of the successful pilot programs developed in several cities.

As the benefits become clearer, more people are willing to support needle exchanges. A 1996 survey by the Kaiser Family Foundation showed that 66% of Americans supported needle-exchange programs. In 1997, the American Medical Association and the U.S. Conference of Mayors urged legislators to revoke the federal law they passed in 1988 prohibiting federally funded needle-exchange programs. The mayors of large U.S. cities, especially those hit hard by the AIDS epidemic, have been strong advocates of these programs, in part because they believe needle-exchange programs are cost effective. They base their belief on evidence that needle-exchange programs slow the spread of HIV, thereby reducing the number of AIDS cases, which results in cost savings for AIDS care. A 2001 report by the U.S. Conference of Mayors on "Best Practices in HIV Prevention and Outreach" highlighted programs in 12 cities: Boston (MA), Cedar Rapids (IA), Chicago (IL), Durham (NC), Houston (TX), Inglewood (NJ), Newport News (VA), San Francisco (CA), Seattle (WA), Trenton (NJ), Tulsa (OK), and Virginia Beach (VA). In several of these cities, the needle-exchange program is one that is featured and described. In addition to explaining how needle-exchange programs operate, the document is also instructive in demonstrating how elected officials can be highlighted as promoters of HIV prevention, including needle-exchange programs. See the web resources for this chapter to view the report.

The number of needle-exchange programs in the United States expanded in the early part of this decade but recently has decreased. In 1997, there were 80 needle-exchange programs in 30 states. In 2002, there were about 200 needle-exchange programs. By 2007, the number had dropped to 185, with programs operating in 36 states, the District of Columbia, and Puerto Rico.

HIV Prevention for Teens

We know that young people are particularly susceptible to HIV. They begin to discover and often to explore their sexuality in their teens, and they are also prone to risk-taking behaviors, perhaps including drug use, as they establish their independence. These actions can put them at risk of HIV transmission.

Long before the advent of AIDS, however, our society has been ambivalent at best about what should be acknowledged, much less done, about adolescent experimentation with sexuality and drugs. On the one hand, teens are confronted with advertisements that implicitly endorse sexuality and drug use (in the form of alcohol and tobacco). On the other, teens are instructed simply to say "no" to sexual contact and drug use. In this battle of explicit and implicit societal mixed messages, HIV/AIDS only add to the confusion and, for some teens, add a new level of tantalizing risk taking, perverse as it may seem to adults.

Rightly or wrongly, adults are the members of society who decide what teens experience or ought to experience and therefore what they do or do not need. Their attitudes about teens are decidedly mixed. Societal attitudes are strongly in favor of protecting teens from risks, especially life-threatening ones like HIV, but other societal attitudes related to teens are not in agreement. Some people have stereotypes of teens as sexual beings who need guidance and instruction. The attitudes of these people support sex education programs, instruction about the risks of HIV, and instruction in HIV prevention methods such as condom use. Other people have stereotypes of teens as individuals who experience sexual matters only if adults raise them and talk about them. To these people, sex education is not only unnecessary but also harmful; HIV risks are not present and therefore HIV prevention instruction is unnecessary. Both groups genuinely believe they understand the situation correctly, and they have the best interests of the teens at heart.

From our understanding of health behavior change (Chapter 9), we know that one critical component for program development is the involvement of the people who are the focus of the program. In the two cases we discussed (Chapter 9, pp. 163–168), the participants themselves—homosexual and bisexual young men in one case and Mexican migrant farm workers in the other—were involved in determining the HIV risks and developing the prevention program. In contrast, what has been striking about the HIV prevention situation with teens is their absence in the discussion of the issue, in the decision about the extent of risk, and in the creation of programs. As long as adults continue to speak for teens, even with the best of intentions, prevention programs aimed at teens will not be effective.

Given societal attitudes, it makes sense to most people that adults should decide whether HIV prevention programs for teens are needed and, if so, create them. It is unlikely, therefore, that teens will be involved in major ways in the creation of HIV prevention programs, although there are cases in which "young people" have been more centrally involved. In these cases, those involved tend to be in their later teen years and early 20s. As a group, these young people come closer to speaking for the needs and concerns of all teens than adults can, thereby providing better input for developing programs and materials.

Healthcare Practices

The story of HIV/AIDS-related discrimination in healthcare practices changed fairly quickly from negative to positive, and it provides an illustration of how attitudes among some groups in society can change, in this case healthcare officials and staff, with more information, data, and knowledge. At the outset of the AIDS epidemic, there were numerous examples of discrimination against those with HIV/AIDS by healthcare workers and institutions. Some workers refused to treat those with AIDS; others did so but with insensitivity and, in some instances, outright hostility. Healthcare workers are not required to treat everyone who comes to them; generally, the law states that treatment cannot be refused in "emergencies." Because of the flexibility in interpretation of who and what should be treated and what constitutes an "emergency," discrimination can play a role in decisions by healthcare workers. It is also important to remember that, when AIDS first arose, there was great uncertainty about what it was and how it could be transmitted. From the perspective of healthcare workers providing direct services to people stricken with this unnamed and unknown disease, caution was prudent because they might become infected too. The confusion and uncertainty at the time does not justify insensitivity, hostility, and discrimination, but it helps us understand how it could happen.

Currently, the situation is greatly improved from that at the outset of the AIDS epidemic, although it is not completely free of problems. A major reason for this is the increase in our knowledge base about HIV/AIDS. Unlike the situation at the outset of the epidemic in the early 1980s, we know how HIV is transmitted and how transmission can be prevented. Also, universal precautions are now in general practice for healthcare workers who can be exposed to potentially infected body fluids, whether the agent is HIV, hepatitis, or other infectious agents. The occurrence of patient-to-caregiver HIV transmission is now extremely rare. Better knowledge and understanding, backed by data, have greatly lessened the fear of HIV infection among healthcare workers and reduced the psychological pressure to take actions that protect healthcare workers at the expense of patients.

The AIDS epidemic has even fostered the development of healthcare approaches relatively new to the United States, such as hospice care for those in the terminal stages of AIDS. Typically, those in the terminal stages of an illness are in a hospital intensive care unit, receiving a variety of medical services, many very costly and of dubious value in prolonging the patient's life. Costs in an intensive care unit can be in excess of $2,000 a day, depending on the care required. *Hospice* is an alternative for those in the final stages of a terminal illness. Hospice care, first developed in England, primarily involves the management of pain so the patient can live comfortably in the final stage of life. Hospice care (where daily costs are much less, in the $250 range, depending on the type and amount of care required) is typically given in the patient's home or a home-like facility and involves a team of caregivers, including those with medical, social, and, if appropriate, spiritual expertise. The focus of attention is not just the patient but also his or her closest family and friends, who themselves become active caregivers. Hospice existed before HIV/AIDS, but the AIDS crisis provided a catalyst for its development as a more humane and cost-effective approach to care during the terminal stage.

At this point, following the discussions in this chapter and the previous one, we have expanded the HIV/AIDS picture as widely as possible, looking at it from the community and society levels. This perspective allows us to see that some societal attitudes can complicate and worsen the HIV/AIDS situation and others can improve it, not only for those with HIV/AIDS but also for everyone else. We know that changing individuals' attitudes and behaviors is very difficult; changing societies' attitudes and behaviors is even more difficult. When societal attitudes begin to shift, however, the effects can be significant. Unlike the situation at the start of the 1990s, when many people did not see HIV/AIDS as a personal concern and some countries even denied that it existed at all, HIV/AIDS is now acknowledged to exist throughout the world. Greater public awareness and understanding have been demonstrated and fostered by large-scale public events such as AIDS walks and the display of the AIDS quilt (Figure 11-2), consisting of memorial panels about those who have died from HIV/AIDS. More important, societies recognize that steps can be taken to prevent HIV infection, and groups like the World Health Organization are assisting in funding and implementing programs throughout the world. These positive societal attitudes and actions are changing the negative actions prompted by old societal prejudices.

Figure 11-2   The AIDS quilt displayed on the grounds of the Washington Mall between the Washington Monument (pictured) and the Capitol Building. (© Hisham Ibrahim/Corbis.)

CHAPTER 12

Future Directions in Combating AIDS

Future Directions for Biomedical Efforts
Prevention of Infection
Treatment of Infected Individuals

Future Directions for Social Efforts
Education
Research

A Final Note of Optimism: Time Is on Our Side

In this book, we have learned about both basic biomedical and social aspects of AIDS. In terms of the biomedical picture, we have considered the virus (HIV), the immune system, the physical manifestations of AIDS, and the transmission of the virus. From the social perspective, we have explored individual risk assessment, the prevention of HIV transmission, and the human and societal aspects of AIDS. However, the fact remains that HIV infection continues to spread in many areas of the world, and there is currently no cure for the disease. How can we respond to this disease, and in what areas are we likely to see activity and progress?

Future Directions for Biomedical Efforts

The biomedical community is focusing on two major problems regarding AIDS: (1) *prevention* of infection through vaccines and (2) *treatment* of infected individuals who develop symptoms of the disease. Let us look at some of the areas where current and future efforts are likely to focus.

Prevention of Infection

Research
Remarkable progress has been made through biomedical research on AIDS in finding and studying the virus itself. The lack of a convenient animal model system is

currently a major stumbling block. Faster progress could be made in understanding the disease process and testing vaccines and therapies if HIV caused a similar disease in experimental animals. However, HIV infects only humans and higher apes, such as chimpanzees; furthermore, the virus does not readily cause disease in chimpanzees. Several retroviruses similar to HIV have been found in monkeys (SIVs, Chapter 4, p. 66), and one strain induces immunodeficiency in rhesus monkeys, so this has been a useful model system. However, monkeys are very expensive to maintain in laboratories, they are in short supply, and the use of primates in research is strongly opposed by some animal welfare advocates. Thus, other more convenient animal model systems are desirable. One possibility is cats: Two retroviruses of cats cause immunodeficiencies. One of these cat viruses (feline immunodeficiency virus) is a lentivirus.

Another recent development has been production of hybrid *simian–human immunodeficiency viruses* (*SHIVs*) by gene cloning techniques. Because SIVs and HIVs are closely related, substitution of HIV genes into SIV often results in a hybrid virus that can still replicate. In particular, SHIVs that consist of the HIV-1 envelope gene inserted into an SIV in place of its own envelope gene can both replicate and cause AIDS in monkeys. These viruses are useful for studying the host immune response to HIV envelope protein during disease development. This kind of SHIV may be valuable for developing and testing anti-HIV vaccines. Other SHIVs may also be useful. For instance, an SHIV containing the HIV reverse transcriptase might be used to test the effectiveness in monkeys of antiviral drugs targeted at HIV reverse transcriptase.

It is also possible to grow cells of the human immune system in special mice. These mice carry a genetic defect called *severe combined immunodeficiency* (*SCID*), which leaves them with crippled immune systems—much like those in AIDS patients. Because SCID mice lack functional cellular immunity, it is possible to implant them with human cells without tissue rejection taking place. Researchers have developed techniques to implant human fetal tissues containing stem cells for the blood into SCID mice. It is then possible to reconstitute these mice with functional human immune cells, including T-lymphocytes and B-lymphocytes. They have also found that if these SCID mice are infected by HIV, the virus will establish infection in the human tissue and destroy the T_{helper} lymphocytes, just as it does in humans. Thus, some of the mechanisms by which HIV attacks the immune system can be studied in these mice. In addition, they may be useful for testing potential antiviral drugs.

Vaccines

Ideally, the most effective prevention of HIV infection would be a *vaccine* that blocks virus infection in an individual. Indeed, effective vaccines have been developed against most human viruses that cause serious diseases (for instance, smallpox, polio, measles, and influenza). Although several possible vaccines against HIV are under development, there are some theoretical reasons that it may be difficult to develop an effective one. As discussed in Chapter 4, HIV evades the immune system in an infected individual. Briefly, this results from (1) the high mutation rate of the virus, particularly in the *env* gene; (2) the ability of the virus to establish a latent state in some cells; and (3) the ability of the virus to spread by cell-to-cell contact. The objective of a vaccine is to raise a protective immune response to the infectious agent. Because HIV evades the immune system so efficiently, it may be difficult for a vaccine to prevent HIV infection in an individual even if it can induce production of neutralizing antibodies or cell-mediated immunity. Another challenge for HIV vaccines is the differences between HIVs found in various parts of the world (see Chapter 6, p. 114). It will probably be necessary to tailor HIV vaccines to the targeted geographical areas.

Despite these theoretical concerns, a number of HIV vaccines have been under development. Most of them have been developed by state-of-the-art gene splicing (or recombinant DNA) techniques that have allowed large-scale production of individual viral proteins. The predominant HIV proteins that make up these potential vaccines are env proteins (e.g., gp120) and, to a lesser extent, gag proteins. Most of these vaccines can raise anti-HIV antibody responses when injected into monkeys or humans. However, the gp120 vaccine failed to provide protection from HIV infection in clinical trials, both in the United States and in Thailand.

Considerable HIV vaccine research has focused on SIVs because they are closely related to HIV, and some SIV strains cause AIDS in certain monkey species. As mentioned earlier, chimpanzees are the only monkeys that HIV can infect, but the virus does not readily cause AIDS in these animals. Thus, many principles of HIV vaccines are being studied by developing analogous vaccines for SIV and testing their abilities to inhibit both SIV infection and disease in rhesus monkeys. A vaccine consisting of killed SIV virus particles has also been prepared. When this killed virus was used to immunize monkeys, it prevented them from developing viral infection or immunodeficiency when injected with low doses of live SIV virus. This was encouraging, but this vaccine protected only against low virus doses. Moreover, it could not protect from infection if live SIV-infected cells were injected—a situation probably closer to natural routes of HIV infection.

As we discussed in Chapter 3, the two branches of the immune system are the humoral immune system, which produces antibody molecules, and the cellular immune system, which produces antigen-specific T-lymphocytes. Because of the intricacies of the immune system, most of the anti-HIV vaccines originally tested were likely to induce anti-HIV antibody responses. However, if cellular immunity to HIV is important for resistance to HIV infection, these vaccines may not be effective. Vaccines designed to induce cellular immunity to HIV are under development as well. Several large-scale trials underway use combinations of vaccines designed to induce cellular immunity with ones that induce humoral immunity.

Due to the nature of the immune system, live viruses are more likely to induce cellular immunity than are killed viruses or pure proteins. Thus, a common approach to vaccines is to use *attenuated viruses*—versions of a virus that can infect an individual without causing disease. The Sabin poliovirus vaccine is an example of an attenuated vaccine, as is the smallpox vaccine. Also, some attenuated SIVs have been shown to act as vaccines in laboratory settings, in that they effectively prevent infection with disease-causing versions of the same SIV in monkeys. However, use of an attenuated HIV as a vaccine is not being currently pursued because of safety concerns—if the attenuated HIV were to mutate in the vaccinated person, it could potentially cause AIDS. On the other hand, genetic engineering approaches are being used to introduce HIV genes into attenuated vaccine strains for other viral infections—for instance, the attenuated viruses that make up vaccines for smallpox and related diseases. These hybrid vaccine viruses will not cause AIDS because they do not express all of the proteins necessary for replication of HIV. On the other hand, they can induce immunity to the HIV proteins specified by the genes that they carry. Numerous versions of such hybrid vaccine viruses are being tested for effectiveness.

In all cases, the first steps for vaccine trials will simply determine whether individuals injected with the test vaccines produce antibodies (or other immune responses) against HIV and if they experience no other harmful side effects (phase I clinical trials; see p. 206). In the initial trials, the proper doses of vaccine to produce an immune response are also determined. Once this has been established, other large-scale trials will test whether the vaccines are effective in preventing HIV infection. Some of the vaccines are currently moving through the vital clinical trial phases.

Testing an HIV vaccine in humans brings together scientific and societal issues, as discussed in Chapter 10. Because of the bioethical issues involved in using humans, groups of specialists in both the biomedical and psychological aspects of HIV/AIDS are working together to implement trials that take scientific, human, and societal considerations into account. There are several national and international networks of researchers working on developing HIV vaccines. These include the HIV Vaccine Trials Network, the International AIDS Vaccine Initiative (IAVI), and the Center for HIV-AIDS Immunology (CHAVI), supported by the U.S. National Institutes of Health. Most recently (2005–2006), the private Gates Foundation has committed $500 million to fund collaborative groups of researchers in HIV vaccine research.

In 2009, the first glimmer of an effective vaccine was reported from a clinical trial conducted by the U.S. Army in Thailand. The approach was to use a live attenuated virus related to the smallpox vaccine virus that expressed HIV protein. This vaccine was followed by a "booster" with a vaccine consisting of HIV envelope protein. While neither of these vaccines showed efficacy when administered by themselves, modest prevention of infection may have occurred, although the analysis of the data is still in progress. If this trial is confirmed, it would establish that a vaccine against HIV can be developed, even if this particular combination itself is not effective enough for use.

Treatment of Infected Individuals

Biomedical efforts to treat HIV-infected individuals focus on three main areas: (1) *antivirals* that interfere with continued HIV infection, (2) *restoration of the immune system*, and (3) *treatments of opportunistic infections and cancers.*

Antivirals

As described in Chapters 4 and 5, several antiviral compounds are now approved for use in treatment of HIV-infected individuals and AIDS patients. These drugs are targeted against three of the HIV proteins: *reverse transcriptase* (nucleoside analogs such as AZT and non-nucleoside reverse transcriptase inhibitors), *protease* (protease inhibitors), and *integrase* (integrase inhibitors). Because these drugs work, agents that interfere with continued HIV infection in an AIDS patient will improve his or her clinical status. Thus, continued efforts are needed to develop new antiviral compounds that also block HIV infection. This is particularly important because drug-resistant HIV frequently appears in infected individuals who are taking an antiviral drug (see Chapter 5, p. 89).

So far, all approved anti-HIV drugs work by interfering with processes carried out by the three viral enzymes reverse transcriptase, protease, and integrase. Other potential antivirals could attack other "Achilles heels" of the virus—processes that are vital to the virus but are not necessary for the survival of the host cell. Nine different genes carried by HIV specify proteins necessary for the virus's life cycle (Chapter 4). Any of these viral proteins are potential targets for new antiviral drugs. Some of the next targets for antiviral drug development are the tat and rev regulatory proteins.

Many additional steps lie between identification of a potential antiviral compound and establishing it as an effective and approved drug. Some of the questions that must be addressed during development of a drug are as follows:

1. What methods are required to deliver effective doses of the compound into individuals?
2. Are there side effects (toxicity), and can effective doses be delivered without side effects?
3. Does the compound inhibit HIV replication in humans?
4. Is the compound as effective as or more effective than the currently available drugs?

Satisfying these questions takes an enormous amount of effort on the part of biomedical researchers and drug companies, as well as a great deal of time. Development of a single antiviral drug requires many millions of dollars and typically 5 to 10 years. In fact, many compounds that show antiviral activity in the laboratory ultimately will never be usable as anti-HIV drugs. Nevertheless, in some cases a compound that shows some antiviral activity in the laboratory might be modified by pharmaceutical chemists into a compound with potency in HIV-infected people; such a compound is referred to as a *lead compound*. Not surprisingly, the high cost associated with developing a new antiviral drug is reflected in the final cost of the drug to the user.

Once laboratory experiments have developed a potential antiviral compound to the point where it shows promise as an anti-HIV drug, clinical trials in humans are conducted. *Clinical trials* are typically conducted in the following sequence:

- *Phase I* clinical trials are conducted on a small number of individuals, and they simply determine the safety of the compound (toxicity or side effects) and establish the methods needed to deliver useful concentrations of the drug into the body. *Phase I trials do not test for effectiveness of the compound.* They are often conducted on uninfected individuals.
- *Phase II* clinical trials are limited in size and test the compound's effectiveness (efficacy). In the case of trials for antiviral drugs, efficacy may be measured by a reduction in symptoms (see Table 6-5), a reduction in laboratory measures of viral infection, or an increase in measures of immune function (discussed in the next section). Phase II trials are conducted in HIV-infected individuals.
- *Phase III* clinical trials are large-scale efficacy trials of a compound. They are typically conducted in multiple locations, and the efficacy of the compound is compared with the efficacy of the currently available therapies. Compounds must show efficacy in Phase III clinical trials before they can receive approval by the U.S. Food and Drug Administration for prescription as an antiviral drug. Before that time, the compounds are considered *experimental* drugs.

Biomedical researchers devote a great deal of effort to developing new antiviral drugs, but it is also important to improve existing drugs. In particular, improvements in the existing formulations of the drugs are beneficial. As described in Chapter 5, taking the current combination antiviral therapies for HIV can be very complicated. Adhering to the strict routines of the combination therapies becomes a major part of the daily existence of HIV-infected individuals. As improved formulations of the antiviral drugs are developed (for instance, development of time-release capsules that decrease the number of times a drug must be taken), they will improve the quality of patients' lives. An example of such an improvement is the ART pill that needs to be taken only once daily (Atripla, see Chapter 5).

As described in Chapter 5, the development of the protease inhibitors and their inclusion in combination therapies in 1996 sparked a great deal of optimism. Because, in some individuals, combination therapies have reduced the viral RNA loads in the blood below the levels of detection, some scientists thought it might be possible to completely eradicate HIV infection. If this were to occur, then the individual could stop taking antiviral drugs. However, as described in Chapter 4, HIV can establish latent infections, and latently infected cells would not be killed by antiviral drugs. Thus, HIV infection is likely lifelong in most infected individuals, even if they are taking combination therapies. It will therefore be important to improve therapies that control HIV infection over extended periods of time.

Another potential class of antivirals is those that interfere with the ability of the virus to enter cells. If the virus entry process is inhibited, then spread of infection within an individual might be inhibited. Recently, a new class of drugs has been developed that blocks HIV infection after virus particle binding to the cell by inhibiting entry of the virus core into the cell. One of these inhibitors, Fuzeon (or T-20, referred to as a "fusion inhibitor"), is now approved for use.

As also discussed in Chapter 4, cells require a co-receptor in addition to CD4 protein for binding and infection by HIV. Therefore, scientists have also focused on blocking the co-receptor interaction with HIV as a way of preventing infection. There is reason to believe that this approach may be effective. A small percentage of the human population completely lacks the CCR5 co-receptor (involved in infection of macrophages) because of a naturally occurring genetic mutation. Epidemiological studies have found that these individuals are resistant to HIV infection after repeated exposures through sexual contact. Thus, an antiviral drug targeted at the CCR5 co-receptor may also be effective in preventing spread of HIV infection. The first generation of co-receptor blocking compounds is in clinical trials.

Three new classes of potential antiviral agents have recently been developed out of basic molecular biology research. One class of compounds is called *antisense molecules*. These are small pieces of single-stranded DNA or RNA that can specifically form double-stranded complexes with HIV viral RNA, similar in structure to double-stranded DNA. Formation of these double-stranded complexes can lead to destruction of the viral RNA. As a result, an infected cell cannot produce viral RNA, viral protein, or virus particles. Current research is focused on establishing methods to effectively deliver these antisense molecules to infected cells and to determine which antisense molecules (directed against which regions of the viral RNA) are most effective. The second class of potential antiviral compounds is called *ribozymes*. Ribozymes are very specialized antisense RNA molecules. When they combine with HIV RNA, they attack the HIV RNA and cut it at particular sites. Thus, they can inactivate virus expression. A third kind of RNA-based antiviral recently under development is called *inhibitory RNA* or *RNAi*. RNAi has some features in common with antisense RNA, but it may be more specific and effective.

As time passes, new potential antivirals will continually appear—some from pharmaceutical laboratories, as discussed, and some from nontraditional sources. In the case of compounds from nontraditional sources (e.g., herbal medicines), they often initially spark great interest, typically based on anecdotal reports of effectiveness. It is important to subject these compounds to rigorous scientific testing (clinical trials) to determine whether they work as claimed. If not, they could worsen the conditions of HIV/AIDS patients who abandon traditional and proven therapies in favor of the new compounds. In the past, one compound that attracted considerable interest was GLQ223, or compound Q. This drug, derived from a Chinese herbal medicine (from bitter cucumber), was found to kill HIV-infected cells in culture. However, standard clinical trials of GLQ223 did not show effectiveness. In addition, serious neurologic side effects, including coma, occurred in some individuals taking GLQ223 in the clinical trials.

Restoration of the Immune System

Most of the clinical symptoms in AIDS result from failure of the immune system, due to depletion of T_{helper} lymphocytes. If the immunological defects can be repaired, then the disease might be arrested or even reversed. As discussed in Chapter 3, all cells of the blood (including those of the immune system) arise by division and differentiation from stem cells that are located in the bone marrow. This process is controlled by a complex series of growth factors that circulate in the body, as described in Chapter 3. Blood cell growth factors are currently the subject of a great deal of research. They are important in many other diseases in addition to AIDS. Ultimately, it may be possible to use these growth factors to stimulate and regenerate the immune system in AIDS patients. Of course, it will be important to use these growth factors in conjunction with antivirals. Otherwise, continued HIV infection would destroy the immune system again. Another potential complication is that growth factors may directly or indirectly activate HIV from latently infected cells.

One growth factor that has shown promise in restoring the immune system in HIV-infected individuals is IL-2 (Chapter 3, p. 36). This is logical because IL-2 is required for growth of both T_{helper} and T_{killer} lymphocytes. A clinical trial on HIV-infected individuals with low T_{helper} lymphocyte counts showed substantial increases in T_{helper} cells after a series of intravenous IL-2 infusions. During this trial, the patients were also treated with antiviral compounds to prevent spread of any latent HIV that was reactivated by the IL-2. However, there are two major drawbacks to this potential treatment. First, IL-2 is an extremely powerful biological molecule. When administered intravenously, it can cause severe side effects, including very high fevers and shock. Patients often receive IL-2 treatment in a hospital setting to manage the side effects. Second, IL-2 therapy will be very expensive: IL-2 itself is very expensive, and hospitalization costs add to the overall cost of the therapy. It would be difficult to provide the current IL-2 therapy to large numbers of HIV-infected people. However, the positive results obtained with IL-2 provide encouragement for developing other, more economical and manageable ways to stimulate the immune systems of HIV-infected individuals. In addition to naturally occurring growth factors for the immune system, several artificial substances that may be able to stimulate immune system regeneration are being developed and tested.

Another possible approach to restoring the immune system would be to supply an AIDS patient with functional T-lymphocytes. Technically, this is very difficult to accomplish because mature T-lymphocytes do not divide. Instead, as described in Chapter 3, it would be necessary to provide new blood stem cells that can divide and differentiate into functional T-lymphocytes. The most logical way to supply these stem cells is through a bone marrow transplant, in which uninfected bone marrow cells are implanted into the recipient individual. These bone marrow cells could then produce functional T-lymphocytes. The greatest technical problem with this approach is that HIV in the infected individual can infect the transplanted bone marrow and destroy the resulting T-lymphocytes. Current cutting-edge research is focused on developing ways to make bone marrow cells resistant to HIV before transplanting them—for instance, by implanting them with an anti-HIV ribozyme.

Currently, there is less emphasis on restoring the immune system than on developing new antivirals. First, experience with ART therapies over the past 10 years has shown that if HIV replication can be blocked, the immune system can substantially restore itself, even in individuals with severely depleted T_{helper} cells. Second, as described, the current approaches to immune system restoration are difficult and expensive.

Treatment of Opportunistic Infections and Cancers

The major practical problems for AIDS patients generally are the opportunistic infections and cancers that result from the lack of immunological protection. Thus, development of better therapies for these opportunistic infections and cancers will play an important role in improved treatment of AIDS patients.

In terms of opportunistic infections, it will be necessary to develop effective drugs for each different opportunistic infection. Many of these infections were rather rare before the AIDS epidemic because the causative agents typically do not cause disease in healthy individuals. As a result, little effort had been put into developing drugs for them. For example, at present no effective treatment can control cryptosporidiosis as an opportunistic infection. The only recourse right now is to treat the symptom (diarrhea). Continued efforts need to focus on developing drugs for these opportunistic infections.

The cancers that result from HIV infection range from Kaposi's sarcoma to tumors of the immune system, called lymphomas. These cancers are quite distinct diseases, and different therapies are necessary for each of them. In the case of Kaposi's sarcoma, one treatment involves use of a naturally occurring protein called α-interferon. Cancer researchers may also provide new therapies for the cancers associated with AIDS. A compound that has recently been used for treatment of Kaposi's sarcoma is Taxol. Taxol was first used in the treatment of ovarian cancer.

Modifying the Conduct of Clinical Trials

Treating HIV-infected individuals often involves new or experimental drug therapies. The AIDS crisis has led to some modifications in the typical clinical trial procedures used for licensing drugs in the United States. As described earlier, the U.S. Food and Drug Administration oversees drug licensing and requires extensive testing in laboratory animals and humans before a drug is approved for therapy. This is a very time-consuming and expensive process, typically taking many years. Largely due to pressure from AIDS activist groups (see Chapter 10), several modifications in these procedures have been developed to speed drug testing and also to make experimental drugs available to patients during the approval process. Traditionally, experimental drugs are administered only through official clinical trials, which usually take place in university research hospitals. To expand the availability of these trials to more patients, *community-based trials* have been established in which experimental drugs are administered to AIDS patients by their local physicians. These physicians then report the results of the treatment to a central source, where the results are pooled.

Another problem with standard clinical trials is that some individuals are too far away from a research university to participate in a trial. As a result, *parallel track* procedures have been developed, in which an experimental drug is made available to a patient (through his or her doctor) in parallel with a clinical trial if that patient is unable to obtain access to the drug otherwise.

Another way AIDS clinical trials have been modified is through the development of *surrogate endpoints*. In clinical trials, the standard yardsticks (endpoints) used to judge a drug's effectiveness are development of clinical disease or death. However, this presents a problem for HIV/AIDS because the time course of infection and disease is so long. When disease or death are used as the endpoints, a clinical trial of an AIDS drug could take many years. As a result, other measurements of an individual's immune system or health have been substituted in preliminary evaluations of HIV drugs. The most common surrogate endpoints are a patient's CD4 (T_{helper}) lymphocyte count, the amount of viral protein (p24 antigen) detectable in the blood, and the viral RNA load in the blood. If a drug lowers the amount of circulating viral RNA or protein or if it increases the T_{helper} lymphocyte count, it would be provisionally considered effective. In fact, the decisions to approve ddI, ddC, and the protease inhibitors as anti-HIV drugs were partly based on clinical trials using surrogate endpoints.

Future Directions for Social Efforts

Infectious diseases do not affect only isolated people. On the individual level, the spread of an infectious agent is caused by the interactions of individuals within a society. On the community level, everyone is directly or indirectly living with HIV/AIDS. Therefore, in combating infectious diseases, it is important to consider society as a whole in planning solutions. This is particularly important for diseases such as AIDS, for which there is currently no cure or vaccine. Social efforts related to AIDS can make contributions in two main areas: education and research.

Education

Two aspects of education can have significant effects on different parts of the AIDS epidemics: education for prevention and education for understanding and compassion.

Education for Prevention

Educational programs targeted to members of high-risk groups are extremely important. These programs are the keys to making these individuals aware of the dangers they face and to promoting changes in behavior that will lessen the risks. As described in Chapter 2, the experience with the syphilis epidemic early in the twentieth century shows the effectiveness of proper public health measures. As also discussed in Chapter 2, public health measures effectively limited the last plague outbreak at the turn of this century—even at a time when there was no cure for the disease. This is analogous to our present situation with HIV/AIDS. As discussed in Chapters 8 and 9, however, changing individual attitudes and behaviors is a challenging task, and, as explained in Chapters 10 and 11, changing societal attitudes and behavior is an equally challenging one. Nonetheless, we are making progress, at both the individual and the societal levels.

In the context of HIV/AIDS, public health education has made an impact. As discussed in Chapter 7, safer sex recommendations have been developed and disseminated, and these have reduced the risk of spreading HIV infection through sexual relations when they have been part of prevention programs. As explained in Chapter 9, for prevention programs to be effective, it is important to work closely with those to whom the projects are targeted. With the spread of the epidemic to minorities and to women, special education programs need to be developed in collaboration with each of these populations.

Continuing to address HIV infection among injection drug users is a critical issue because these individuals are one conduit for the spread of HIV infection into the general heterosexual population. The current programs, such as needle-exchange programs, have been successful but are not as widespread as they should be to address this challenge. Also, they are underfunded. For reasons described in Chapter 11, it is hard, slow work to gain public acceptance for public health measures and programs targeted to injection drug users. Such programs have been opposed by people who argue that distribution of needles condones and encourages drug addiction. Although some visible public figures, including city mayors and state governors, have spoken up in their support, needle-exchange programs are having a challenging time maintaining their services, and the number of new programs is not increasing in the United States.

Another important goal of public health education is to prevent backsliding in behavior. Behavior modification through public education has clearly been effective in areas where the AIDS epidemic has hit hard—for instance, the homosexual male community in San Francisco. However, recent follow-up studies have detected a significant frequency of reversion to high-risk sex practices by some men in this community as time has passed. We could have predicted these reversions based on the precaution-adoption process model discussed in Chapter 9, p. 156. It is important to develop methods to promote continued adherence to safer behaviors, even after initial efforts have been effective. We cannot forget the seventh health promotion/ disease prevention program principle, the scientific principle: All our programs need to be evaluated for their short- and long-term effects.

Education for Understanding and Compassion

Another aspect of education is equally important: educating the general public about HIV, AIDS, and those with the disease. As we discussed in Chapter 10, at the outset of the epidemic, fear and lack of knowledge and experience fostered prejudice and discrimination against those with HIV/AIDS. Education of various types and in various forms helped to lessen the fear of HIV/AIDS by increasing knowledge about what HIV was, how it was spread, and how it was medically treated. As knowledge has grown, it has been easier to develop and implement public policies to decrease discrimination against those with HIV/AIDS and to increase programs to stop the spread of the disease.

Healthcare workers and the healthcare system are other important targets for educational efforts and capacity development. The number of doctors with special expertise in treating HIV-positive individuals and PWAs is growing. Some special clinics or wards in hospitals have made treating PWAs their specialty. In these cases, the quality and sensitivity of care for people living with AIDS has been much better than what used to be generally available. The lessons learned from these settings need to be disseminated to other healthcare workers so that more HIV-positive individuals and PWAs can benefit. The lessons should also be shared with health system administrators so that capacity is built to provide sensitive HIV care.

From the discussion in Chapter 6, we know that more individuals are being infected with HIV, are progressing from HIV-positive status to AIDS status, and are living longer with HIV disease. The increases in numbers of individuals affected by the disease are further straining our current healthcare and social service systems. For the wisest decisions to be made about the allocation of scarce financial and personnel resources, we need to have a general population that is informed about HIV/AIDS. The decisions will be difficult enough without fear and prejudice clouding the public's consideration of options.

Research

All these education efforts rely on good research. Whether we are developing HIV prevention programs for individuals or AIDS education programs for the general public, we need to draw on the research and theories available to us. For example, as we discussed in Chapter 9, the development of effective HIV prevention programs requires careful and detailed attention to a number of factors, such as cultural and social aspects of people's situations. Good social research will allow us to identify these factors and determine which approaches will most likely be effective.

We also need to improve our epidemiological and survey research related to AIDS. As HIV/AIDS has moved into new communities, our epidemiological research has been slow to catch up. For example, AIDS is occurring in the Asian/Pacific Islander community in the United States. Initially, epidemiological data grouped these individuals in the "other" category, instead of the more frequently chosen racial and ethnic categories (white, Hispanic, African American, etc.). Now, Asian/Pacific Islander is listed separately under racial and ethnic groups and, in some surveys and studies, is broken down into more specific groups. The cultural practices and beliefs are different enough between Asian/Pacific Islander subgroups that, for effective prevention planning, we need to know more about exactly which subgroups are most affected by HIV.

Likewise, we need accurate survey research on HIV- and AIDS-related knowledge, attitudes, and practices among different segments of the population. We have thorough survey research on certain special study groups of PWAs (for instance, homosexual men), which has been invaluable in planning and implementing both biomedical and psychosocial programs. Our research on other segments of the population affected by HIV is less complete or missing. For example, we have very little or almost no systematic research on certain subgroups (such as sex workers) and HIV. Without good research, we are unable to develop a thorough understanding of the HIV-related context and the individuals and factors within it.

Finally, we need better evaluation of all our HIV/AIDS programs. Only through scientific research can we be sure that we are implementing programs with beneficial effects. Too often, we establish a program with the best of intentions but without a sound scientific plan to assess the actual effects or to detect unintended consequences, both positive and negative. Scientific evaluation provides those working to address the social aspects of AIDS with a way to know whether their efforts are effective. Decreases in HIV infection rates in particular subpopulation groups are the ultimate outcome of many prevention programs. Unfortunately, we cannot always measure this outcome;

even when we can, HIV infection rates focus on groups and therefore change slowly over a long period of time. For the most complete picture, we need to assess changes in the entire chain, including knowledge, attitudes, intentions, practices, and HIV status.

A Final Note of Optimism: Time Is on Our Side

Many of the facts and statistics about AIDS in this book are frightening and depressing, especially because a cure has not been developed yet. Indeed, those who are suffering from the disease or are at risk to develop it often express frustration at the apparent lack of progress in AIDS research. But let us look at some time scales to get a sense of perspective. First, as discussed in Chapters 4 and 6, the current estimates are that most HIV-infected individuals will develop AIDS, with an average time between initial infection and disease symptoms of 10 to 12 years, even if they do not receive antiviral drugs. Thus, new therapies and treatments that are developed in the next 5 or 10 years may help many of those who are currently infected.

Second, let us look at the rate of scientific progress in the AIDS epidemic. For comparison, let us consider two other diseases that have had great impacts on society: the ancient disease, plague (black death), and the more recent disease, polio (see Chapter 2). Table 12-1 compares the time scales for fighting these diseases. Plague probably first caused major epidemics as early as the fifth or sixth century; the well-documented black deaths occurred in the fourteenth and following centuries. The infectious agent, *Yersinia pestis*, was finally isolated in 1894. Effective therapy against the disease had to wait for the development of classical antibiotics in the 1940s.

Polio was first recognized as an epidemic disease in the 1880s, and the infectious agent, poliovirus, was isolated in the late 1940s. Even after the virus was identified, there was no effective therapy for individuals once they became infected. Ultimately, the disease was brought under control by the development of the Salk and Sabin polio vaccines beginning in 1955.

Table 12-1	A Time Comparison of Three Epidemics		
Disease	First Documented Epidemic	Isolation of Agent	Therapy
Plague (*Yersinia pestis*)	A.D. 560	1894	1940s (antibiotics)
Polio (poliovirus)	A.D. 1885	1909 identified 1949 isolated	1953 (Salk vaccine)
AIDS (HIV)	A.D. 1981	1984	1986 (AZT partially effective)

As for AIDS, the disease was first recognized in 1981 and the causative agent, HIV, was isolated in 1983–1984. By the end of 1986, the first partially effective antiviral, AZT, was developed; it was put into wide use in 1987. The protease inhibitors and triple combination therapies were introduced in 1996, and they provide a significant improvement in the management of HIV-infected individuals. Indeed, the introduction of the combination therapies, or ART, in 1996 has led to a decrease in the rate of death from AIDS in those countries that have access to these drugs. For instance, as shown in Table 12-2 and Figure 12-1, in the United States there was a 45% decrease in deaths from AIDS in the first half of 1997 as compared with the first half of 1996 (the triple

Table 12-2	Aids Cases and Deaths in the USA: 1996 vs. 1997*		
	January–June 1996	January–June 1997	Decrease 1997–1996 (%)
Total AIDS cases	33,243	28,370	15
Total AIDS deaths	21,281	11,479	45

*Data from CDC, *HIV/AIDS Surveillance Reports*. Protease inhibitors and the triple combination (HAART) therapies were introduced in mid-1996, so January–June 1996 represents the last period before introduction of these therapies.

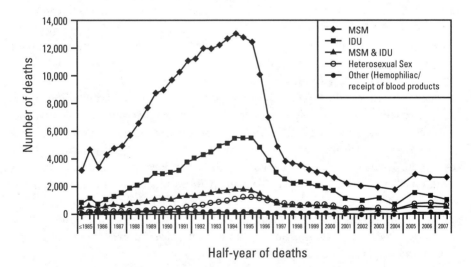

Half-year of deaths

Figure 12-1   Estimated adult/adolescent deaths among men with AIDS, by risk exposure and half-year of death, 1985–2007, United States. MSM = men who have sex with men; IDU = injection drug use. (Reproduced from CDC, *HIV/AIDS Surveillance Supplemental Report 14* [2006]: Figure 7.)

combinations became available in mid-1996). (In this case, it is important to remember that this does not mean that the number of HIV-infected people has decreased or that the rate of new HIV infections has decreased; the decline is related to the ability of the triple combination therapies to inhibit progression of HIV-infected people to AIDS.) However, the rate of decrease in AIDS deaths slowed in subsequent years (Figure 12-1), reflecting the limitations of ART (see Chapter 5), and it has now leveled off. Thus, the rate of progress in AIDS research has actually been very rapid in historic terms, although new approaches are needed to further reduce the rate of AIDS deaths. It is also important to remember that the decrease in AIDS deaths in more developed countries reflects the fact that most HIV-infected people have access to antiviral drugs. In other parts of the world where HIV-infected individuals do not have ready access to these drugs (and where the most HIV infection is), these improvements in death rates have not occurred.

The progress in AIDS biomedical research largely reflects the great advances in molecular biology, virology, immunology, and biotechnology that have taken place over the last 40 years. For instance, the life cycle of retroviruses was worked out largely in the 1970s and the 1980s, after the discovery of reverse transcriptase. In terms of immunology, the understanding of the different kinds of lymphocytes (B versus T; T_{killer} versus T_{helper}) is also recent. The techniques to identify the CD4 protein of T_{helper} lymphocytes are less than 30 years old. It is difficult to imagine how much more serious the AIDS epidemic would be if it had struck 35 years ago, before these advances. One program that provided a major boost to these fields was the War on Cancer, which was launched by the U.S. government in 1971 to conquer cancer with the same approach used to put humans on the moon. Although the War on Cancer has not been won yet, the program resulted in a great deal of research on retroviruses, and it heavily contributed to the development of recombinant DNA cloning technologies. This has been essential to the rapid achievements in AIDS research. Because biomedical research has advanced so rapidly in the last few years, we are optimistic that new and more effective solutions to HIV/AIDS will be developed in the not-too-distant future.

In personal terms, we can take positive approaches to dealing with the AIDS epidemic. Because death is an inevitable part of our lives, it is more productive to focus on wellness and the quality of life than on illness and death. This applies to both those who have AIDS and those who do not—as well as to people affected by cancer or other terminal illnesses. Like all major social changes, AIDS presents us with not only problems but also opportunities on both biomedical and social levels. For example, we have already expanded our scientific knowledge of the immune system due to the efforts to understand AIDS. We also have a better understanding of the important factors in successful disease prevention and health promotion programs. There are many other opportunities to make progress on biological and social issues. AIDS is a crisis and an opportunity for social improvement: The challenge is to use the opportunities for greater personal, social, and biological understanding.

http://biology.jbpub.com/fan/aids/6e/

Connect to this book's website: http://biology.jbpub.com/fan/aids/6e/. The site features summaries of the main points from each chapter, links to important AIDS-related websites, and short-answer-style review questions for each chapter.

Glossary

ACT UP Originally, the acronym for "AIDS Coalition to Unleash Power." Now, locally based, loosely organized associations of PWAs and their supporters.

AIDS Acquired immune deficiency syndrome is an incurable infectious viral disease that results in damage to the immune system in otherwise healthy individuals.

Analytical epidemiology Epidemiological studies that seek to identify and explain the causes of diseases.

Anchoring One judgment heuristic based on the starting point for an assessment and its effect on subsequent assessments.

Anemia A condition of the blood in which there are abnormally low concentrations of red or white blood cells.

Anonymous A term used to describe HIV testing situations in which the test results cannot be linked to an individual's name.

Antibiotics Compounds that are effective against infection by microorganisms such as bacteria, fungi, and protozoa. They are generally ineffective against virus infections.

Antibody A protein produced by a B-lymphocyte that specifically binds a particular antigen. This leads to attack by the immune system.

Antigen A molecule or substance against which a specific immune response is raised.

Antisense molecules DNA or RNA molecules that can form specific double-stranded couples with HIV RNA. They can bind to HIV RNA and prevent it from functioning in the infected cells and are being explored as therapeutics for HIV infection.

Antivirals Compounds that are effective in treating virus infections.

Asymptomatic AIDS carriers Individuals infected with HIV who do not show any sign of disease. They may be capable of infecting others.

Asymptomatic infection An infection for which there are no superficially visible or noticeable changes in the body or its functions that indicate the presence of the infection. Contrast with **Symptomatic disease**.

Attitude An individual's overall evaluation of information regarding people, objects, or issues.

Availability One judgment heuristic based on the presence of an item or object in memory. The two main components of availability are familiarity and salience.

Azidothymidine (AZT) Also called Retrovir or zidovudine. An antiviral that is effective in treating HIV infection and AIDS. It works by preferentially being incorporated by reverse transcriptase into growing viral DNA during HIV replication.

Bacteria Small single-cell microorganisms that can cause diseases.

Behavior An individual's action regarding a specific issue.

B-lymphocytes One kind of lymphocyte. B-lymphocytes secrete antibodies that are specific for particular antigens.

Case/control studies A form of analytical epidemiology in which a group of individuals with a particular disease (the cases) are compared with a matched group of unaffected individuals (the controls).

Case reports Reports and descriptions of an unusual disease occurrence in individual patients. Case reports are one form of descriptive epidemiology.

Causality The factors contributing to the development of disease in epidemiological studies.

CD4 protein A surface protein that is characteristic of T_{helper} lymphocytes. It is also present on some macrophages and dendritic cells. CD4 protein is the cell receptor for HIV.

Cellular immunity Immunity involving T-lymphocytes (particularly T_{killer} lymphocytes).

Circulatory system The system of vessels that moves blood around the body, including arteries, veins, and capillaries.

Clades Subgroups of HIV. Various human populations throughout the world are infected by different clades of HIV. In North America, the predominant clade of HIV-1 is clade B.

Clinical trial Trials of drugs, vaccines, or therapies in humans. They are the final tests used before the treatments are approved for public use.

Cognition A representation in an individual's mind arising from thinking or knowing.

Cognitive dissonance An imbalance in cognitions caused by contradictions between elements of the cognitions. The imbalance creates psychological tension for resolution and the restoration of balance.

Cohort studies A form of analytical epidemiology in which a group of individuals who share a particular risk factor for a disease are studied.

Combination therapies (also triple combination therapies, AIDS drug cocktails, or HAART) Combinations of antivirals administered to HIV-infected individuals. They have been found to be considerably more effective than single antiviral drugs alone in giving sustained reductions in viral RNA load. See also **HAART**.

Confidential A term used to describe HIV testing situations in which the results are linked to an individual's name in an identifiable but protected way.

Control group A group of individuals who serve as the scientific comparison for a similar but separate experimental group of individuals who receive a special treatment or intervention. The presence, meaning, and significance of changes in the experimental group due to the special treatment or intervention are identifiable through comparisons with the control group. See also **Experimental group**.

Co-receptors for HIV Cell surface molecules required along with surface CD4 protein for HIV binding and infection of a cell. The two predominant HIV co-receptors are called CCR5 and CXCR4. CCR5 is the co-receptor found on macrophages, and CXCR4 is the co-receptor on T_{helper} lymphocytes.

Cross-sectional/prevalence studies Monitoring a population for occurrence of diseases and noting the time and kind of disease. A form of descriptive epidemiology.

DC-SIGN A molecule on the surface of dendritic cells that binds HIV particles and allows transport of the virus particles to the lymph nodes.

Dementia Loss of mental function due to damaged brain cells and brain inflammation in AIDS-afflicted patients.

Dendritic cells Cells of the immune system that are in the skin and the surface of internal tissues such as the colon. They carry antigens to lymph nodes where they present the antigens to B- and T-lymphocytes. Dendritic cells have CD4. They also can carry HIV virus particles to T-lymphocytes in the lymph nodes.

Descriptive epidemiology Epidemiological studies that describe the occurrence of disease by person, place, and time. Generally the first kinds of studies carried out in a new disease.

Discrimination Behavior directed toward an individual based only on the individual's membership in a particular group.

Drug-resistant HIV Virus that frequently appears in HIV-infected individuals taking an antiviral drug. Once a drug-resistant HIV appears, then that antiviral drug is no longer effective in that person.

ELISA The most common test for HIV antibodies.

Endemic pattern Patterns of continuous infection that allow epidemic diseases to remain present in populations.

Epidemiology The study of patterns of disease occurrence in populations and of the factors that affect them.

Experimental group A group of individuals who receive a special treatment or intervention and who are compared with a control group of similar but separate individuals. See also **Control group**.

Experimental/interventional studies A form of analytical epidemiology in which a condition in a population is changed, and the effect on disease development is observed.

Fotonovela (Spanish for "photo booklet") A photo story book with pictures or sketches with captions in Spanish. Similar to a comic book in format.

Fungi Microorganisms that may exist as single cells or are organized into simple multicellular organisms.

Germ theory The postulate (1546) that infectious bacterial, fungal, or viral organisms cause disease.

HAART (highly active antiretroviral therapy) A combination of antiretroviral drugs that efficiently inhibit HIV replication in infected individuals. Typically the combinations include two nucleoside inhibitors and one protease inhibitor. These are also called triple combination therapies or AIDS drug cocktails.

Helper T-lymphocytes T-lymphocytes that help T_{killer} and B-lymphocytes respond to antigens. Destruction of T_{helper} lymphocytes is the major problem in AIDS.

Heuristics See **Judgment heuristics**.

HIV (human immunodeficiency virus) The virus that causes AIDS; previously called HTLV-III, LAV, and ARV. The predominant form of HIV in North America, Europe, and central Africa is called HIV-1. A closely related retrovirus found in western Africa is called HIV-2.

HIV antibody test A test to determine whether an individual has antibodies to HIV, the virus that causes AIDS. Presence of HIV-specific antibodies indicates that the person has been exposed to HIV and has raised an immune response, but it does not actually tell if the person is still infected. The most common test is the ELISA test. A backup test called the Western blot is also used.

Hospice An approach to caring for individuals in the terminal (or final) stages of an illness characterized by pain management instead of medical intervention and by attention to issues related to the individual in his/her social context (of family and friends).

Humoral immunity Immunity involving B-lymphocytes and the antibodies they produce.

IDU Injection drug user.

Immune system The circulating cells and serum fluids in the blood that provide continuous protection from foreign infectious agents.

Immunological memory The ability of the immune system to respond rapidly to a previously encountered antigen with specific antibodies.

Incidence The proportion of a population that develops new cases of a disease during a particular time period.

Incubation period The period between infection by a microorganism and appearance of disease symptoms.

Intention An individual's decision to act in a particular manner regarding a specific issue. (Contrast this with **Behavior**.)

Interleukin 2 (IL-2) A growth factor required by T-lymphocytes. It is produced by stimulated T_{helper} lymphocytes and is required by both T_{helper} and $T_{cytotoxic}$ lymphocytes for growth.

Judgment heuristics Individualistic rules of thumb used in subjective probability models of decision making.

Kaposi's sarcoma A normally rare cancer that develops frequently in AIDS patients.

Killer or cytotoxic T-lymphocytes T-lymphocytes that kill the target cells they bind to.

Koch's postulates A series of criteria used to establish that a particular microorganism causes a disease.

Latency A state of virus infection in which the virus's genetic material remains hidden in the cell, but no virus is produced. At a later time, the latent virus may become reactivated. HIV can establish latent infection, particularly in macrophages.

Lead compound A compound in drug development that shows activity against HIV in a laboratory culture. It may be modified by a pharmaceutical chemist to make even more effective antiviral compounds, some of which may eventually be useful in management of HIV infection in people.

Lentiviruses A subclass of retroviruses that includes HIV. Some lentiviruses infect other species, including monkeys, sheep, and cats.

Lymphadenopathy syndrome (LAS) Persistently enlarged lymph nodes or swollen glands, sometimes an early sign of HIV infection that is progressing. Also called persistent generalized lymphadenopathy.

Lymphatic circulation A second circulatory system through which lymphocytes circulate. Lymph channels drain fluid from tissues (lymph) into lymph nodes, where B- and T-lymphocytes are located. Antibodies or T-lymphocytes are produced in the lymph nodes in response to infection, and they enter the general circulation by way of other lymph channels.

Lymphocytes Cells of the immune system that respond specifically to foreign substances. There are several kinds of lymphocytes. The two classes of lymphocytes are B-lymphocytes and T-lymphocytes.

Lymphoma Cancer of lymphocytes of the immune system.

Lytic infection Infection of a cell by a virus that results in death of the cell. HIV infection of T_{helper} lymphocytes is a lytic process.

Macrophages One kind of phagocyte. Macrophages generally attack cells infected with viruses.

Nonlytic infection Infection of a cell by a virus that results in production of virus but survival of the cell. Most retroviruses normally produce nonlytic infections. HIV infection of macrophages is nonlytic.

Non-nucleoside reverse transcriptase inhibitors Anti-HIV drugs targeted to the HIV reverse transcriptase but distinct from the nucleoside analogs. Non-nucleoside reverse transcriptase inhibitors bind directly to the viral reverse transcriptase and directly inhibit its action.

Norm A standard about appropriate attitudes or behaviors for individuals or groups. The standard is socially defined or redefined, and it is maintained through social pressure. Some norms are formalized into laws that are maintained through legal means.

Normative model A decision-making approach that uses probabilistic information according to statistical rules to reach conclusions.

Nucleoside analogs A class of antiviral drugs that inhibit HIV replication. They are incorporated by reverse transcriptase into the growing viral DNA molecule, but once incorporated they prevent further growth of the viral DNA. AZT was the first nucleoside analog approved for treatment of HIV-infected people. Five nucleoside analogs have been approved for treatment in HIV/AIDS: AZT, ddC, ddI, d4T, and 3TC.

Opportunistic infections Infections by common microorganisms that usually do not cause problems in healthy individuals. Opportunistic infections are the major health problems for AIDS patients.

Optimistic bias The tendency of an individual to believe that, compared with others, good things will happen to him or her and bad things will not happen.

Pandemic disease An infectious disease present on many continents simultaneously.

Perceived severity An individual's personal opinion about the severity of a disease. Perceived severity may or may not be closely related to actual severity. See **Severity**.

Perceived susceptibility An individual's personal opinion about his or her susceptibility to a disease. Perceived susceptibility may or may not be closely related to actual susceptibility. See **Susceptibility**.

Personal invulnerability The tendency of some individuals to believe they are not susceptible to harmful risks. See **Optimistic bias**.

Phagocytes Cells of the immune system that eat foreign cells or infected cells. The two kinds of phagocytes are macrophages and neutrophils (granulocytes).

Phase I clinical trials The first clinical trials carried out during development of a drug or a treatment. Phase I clinical trials test for toxicity, and they establish the methods for achieving effective doses of the drug or treatment. They do not test for effectiveness (efficacy).

Phase II clinical trials Limited clinical trials that test for efficacy of the drug or treatment.

Phase III clinical trials Large-scale clinical trials that test for efficacy in multiple settings and compare the efficacy of a drug or treatment with the currently available drugs/treatments.

Polymerase chain reaction (PCR) A sensitive method for detecting the presence of a specific DNA or RNA. PCR-based assays have been developed for detection of HIV viral DNA and RNA. They are the basis for the HIV viral RNA load assays.

Prejudice An unfavorable attitude toward a group of people.

Prevalence The fraction of individuals in a population who have a disease or infection at a particular time.

Primary immune response The immune response that follows the first exposure to an infection or an antigen first. There is a lag period before antibodies are produced.

Probabilistic information Material containing a statistical estimate related to an issue or topic.

Probability A statistical estimate of likelihood, usually expressed as a proportion (from .00 to 1.00) or a percentage (from 0% to 100%).

Protease An enzyme encoded by HIV and other retroviruses. It is important in maturation of the virus particle and is required for infectivity.

Protease inhibitors Antiviral drugs that inhibit HIV protease. They have become important drugs for management of HIV infection, particularly when used in triple combination therapies with nucleoside analogs.

Protozoa Large single-cell microorganisms that can cause diseases.

PWA Person with AIDS.

Quarantine Enforced isolation of individuals with infectious disease or of individuals suspected of having an infectious disease.

Rapid testing HIV antibody tests that are done in one session, with results available at the testing session.

Red blood cells (erythrocytes) Blood cells that are responsible for carrying oxygen to and carbon dioxide from the tissues.

Representativeness One judgment heuristic based on assumed similarity between two objects or items. One object or item is assumed to be representative of another to the extent that the two objects or items are similar.

Reverse transcriptase An enzyme that is unique to all retroviruses. It reads the genetic information of the retrovirus, which is RNA, and makes a DNA copy.

Ribozymes Specialized forms of antisense RNA molecules. They can cause the cutting of HIV RNA in infected cells and are being explored as potential therapeutic molecules for HIV infection.

Risk assessment An evaluation of the susceptibility of a group or an individual to a particular threat (e.g., HIV infection).

Role A position in a social setting involving interrelationships between and among people. An example is the role of student, nurse, or patient.

Role expectations The assumptions about the behaviors that an individual in a particular role will exhibit.

Schema A combination of similar or related cognitions. Schemas serve as reference points for organizing an individual's past experiences or interpreting new experiences, or they serve as templates for activating new ideas.

Secondary immune response An immune response that follows exposure to an infection or an antigen that the immune system had encountered before. The strength of the subsequent response is greater, occurs more rapidly, and lasts longer.

Self-concept The schema an individual has about himself or herself. This self-schema is socially constructed and socially maintained.

Self-esteem The positive or negative evaluation component of the self-concept.

Seronegative An individual who tests negative for HIV antibodies.

Seropositive An individual who tests positive for HIV antibodies.

Severity Seriousness of a disease in terms of causing unpleasant or harmful effects on the body, possibly to the point of being life threatening. See also **Perceived severity**.

SHIV Hybrid simian–human immunodeficiency virus. SHIVs are generated in the laboratory and consist of an HIV gene substituted into SIV for the equivalent SIV gene. Useful SHIVs are those that replicate and/or cause disease in monkeys.

Simian immunodeficiency virus (SIV) A group of retroviruses closely related to HIV that are native to old world (African) primates. Some SIVs cause AIDS when infected into particular primate species. SIVs are often used as experimental models for HIV. HIV-1 is most closely related to SIV from chimpanzees, and HIV-2 is most closely related to SIV from sooty mangabeys.

Stereotype Group-based schema generally reflecting society's appraisal of a group. Stereotypes can be either positive or negative.

Stigma A negative assessment associated with a particular object or issue.

Subjective probability model A decision-making approach that uses probabilistic information according to individualistic, personal biases to reach conclusions.

Surrogate endpoint Indicators of the effectiveness of drug therapies, which substitute for the standard indicators (i.e., development of disease or death). For HIV, surrogate endpoints are measures of the condition of an individual's immune system such as CD4 lymphocyte counts, the amount of viral protein (p24 antigen) in the blood, and viral DNA load.

Susceptibility Capacity of a person to be infected by or unresistant to a disease. Contrast with **Perceived susceptibility**.

Symptomatic disease A disease for which there are superficially visible or noticeable changes in the body or its functions that indicate the presence of disease. These changes are unpleasant or harmful and thus call attention to the consequences of the disease. Contrast with **Asymptomatic infection**.

Test group See **Experimental group**.

T-lymphocytes One kind of lymphocyte. Unlike B-lymphocytes, T-lymphocytes do not release antibodies, but they specifically recognize and bind foreign antigens. The two main types of T-lymphocytes are T_{killer} and T_{helper} lymphocytes.

Treatment adherence The ability of a person undergoing therapy to follow the prescribed treatments. Treatment adherence is a serious concern in triple combination therapies of HIV-infected people.

Vaccine A preparation that can induce protective immunity to a microorganism such as a virus or bacterium. Some vaccines are inactivated or attenuated microorganisms, and others consist of purified proteins of the microbe.

Viral envelopes Structures that surround some virus particles and resemble membranes around cells. Viral envelopes contain virus-specific proteins that are important in binding cell receptors. Viral envelope proteins are major targets for the immune system.

Viral RNA load (or viral load) The amount of HIV RNA detectable in the blood of an infected person. This reflects the amount of HIV virus particles in the blood. Viral RNA load measurements use the PCR reaction, which is very sensitive. They are much more sensitive for detecting HIV infection than either ELISA assays for HIV antibodies or assays for viral p24 antigen.

Viruses Small infectious agents. They are parasites that grow only inside cells.

Western blot An HIV antibody test, used as a confirmatory test for a positive ELISA test.

White blood cells (leukocytes) All blood cells except red blood cells, including lymphocytes, neutrophils, eosinophils, macrophages, and megakaryocytes.

Window period For an individual, the period of time between infection by a virus and the production of antibodies to the virus.

Xenophobia Discriminatory fear of foreigners.

APPENDIX

Reference Resources

HIV/AIDS is a continually evolving topic. New biomedical and social developments happen almost daily. This appendix provides the reader additional sources to find the latest information on HIV/AIDS.

Websites

A great number of HIV/AIDS websites can provide the latest scientific, medical, and social updates. If a website does not have the information desired, checking linked websites is recommended. Several well-organized and maintained websites are listed here. Direct links to these sites can be found at this text's website located at http://biology.jbpub.com/fan/aids/6e/.

CDC Division of HIV/AIDS Prevention
http://www.cdc.gov/hiv/dhap.htm
The U.S. Centers for Disease Control and Prevention website, providing the latest epidemiological statistics and downloadable slides on HIV/AIDS.

CDC National Prevention Information Network
http://www.cdcnpin.org/
A Centers for Disease Control and Prevention website focused on prevention of HIV infection.

UNAIDS
http://www.unaids.org/en/
The United Nations AIDS program website, offering an excellent global AIDS perspective.

AIDSinfo
http://www.aidsinfo.nih.gov/
The U.S. National Institutes of Health website, providing up-to-date information on HIV/AIDS treatments and clinical trials.

AIDS Education Global Information System (AEGIS)
http://www.aegis.com/
An extremely comprehensive AIDS website, covering all aspects of AIDS; updated hourly.

New York Times
http://nytimes.com/
A searchable website for the *New York Times*, offering current and past articles related to HIV/AIDS.

Los Angeles Times
http://latimes.com
A searchable website for the *Los Angeles Times*, offering current and past articles related to HIV/AIDS.

Johns Hopkins AIDS Service
http://www.hopkins-aids.edu/
An excellent university-based website, offering up-to-date information on HIV/AIDS as well as other diseases.

Books

Abbas, A. K., A. H. Lichtman, and J. S. Pober. (1997). *Cellular and Molecular Immunology*. Philadelphia, PA: W. B. Saunders.
An advanced/graduate-level text in immunology.
Benjamini, E., G. Sunshine, and S. Leskowitz. (1996). *Immunology—A Short Course*, 3rd ed. New York: Wiley-Liss.
A college-level immunology text.
Coffin, J. M., S. H. Hughes, and H. E. Varmus, eds. (1997). *Retroviruses*. Plainview, NY: Cold Spring Harbor Laboratory Press.
The authoritative graduate/professional-level reference book on retroviruses. It includes extensive discussion of HIV.
Stine, G. (1998). *Acquired Immune Deficiency Syndrome: Biological, Medical, Social, and Legal Issues*, 3rd ed. Englewood Cliffs, NJ: Prentice-Hall.
An entry-level college text on HIV/AIDS; contains many facts.

Index

The abbreviations *t* and *f* stand for table and figure, respectively.